More than Brooms

GIS-based Survey and Mapping of Tiger Grasses in Northern Tablas, Philippines

EDDIE G. FETALVERO

ISBN-13: 978-1502983831
ISBN-10: 1502983834

DEDICATION

To my wife, Sweet and to our two smart kids, Poochi and Sam, this piece work is lovingly dedicated to you for your constant love, care and understanding. By giving it our best and trying hard until we succeed, we are already winners in our own right.

CONTENTS

ACKNOWLEDGMENTS

I am indebted to the following persons who in one way or another shared to me their time, resources and expertise to bring this study into full completion.

To Dr. Josefina P. Abilay, the DOST IV-B Director for the financial assistance extended by her office through the Romblon Provincial Science and Technology Center headed by its supportive coordinator, Dr. Bilshan F. Servañez. This break helped me learn first-hand lessons in R&D management.

To Dr. Jeter S. Sespeñe, the former President of Romblon State University, for keeping me motivated and for believing in my capacity to handle projects like this and to his successor, Dr. Arnulfo F. de Luna for his continued support in the research activities of the University;

To the barangay captains of the villages surveyed: Hon. Agustin M. Magracia of Binongaan, Hon. Pacifico M. Moral of Doña Juana, Hon. Enrico S. Morgado of Hinugusan, Hon. Jovita G. Guro of Mari-Sur, Hon. Robert F. Gabon of Mari-Norte, Hon. Edilberto M. Guntan of Victoria, and Hon. Dominador N. Ogatia of Pagsangahan, for accommodating our team of enumerators and surveyors in their respective barangays and for ensuring their safety all throughout the duration of the survey.

To the people we met in those villages who served as guides as well as those who offered their homes for us to stay: Kagawad Arnold Morada of Hinugusan, Sammy Gadon of Mari-Norte, Ricky Ignacio of Mari-Sur, Ma'am Renita Morada and Jervic Merida of Pagsangahan and most specially to Mrs. Marichu Molina of Doña Juana for making our stay in their barangay memorable.

To Fred Fabila for his assistance in helping us with the basics of GPS and to Sheilla Jane Forlales whose expertise in converting the readings into GIS maps paved the way for the completion if this research work. Likewise, to my research team, surveyors Kit and Wilmar; enumerators, Arlan, Francis, Brenda Lee, Suzanne and Joelyn; and our encoders Jenjen and Maimai, for their commitment to the project and for their willingness to do the extra mile.

And most of all, to God who sustained me with strength, knowledge and wisdom all throughout the study, be all glory and praise!

EXECUTIVE SUMMARY

I. Project Title

MORE THAN BROOMS

GIS-Based Mapping and Survey of Tiger Grasses in Northen Tablas, Philippines

II. Researcher

Eddie G. Fetalvero
Associate Professor III
Romblon State University
Odiongan, Romblon

III. Fund Source

Department of Science and Technology MIMAROPA through its SETUP Project
Romblon State University

IV. Study Duration

January 2012 – March 2013

V. Summary

Tiger grass industry is an economic activity in the uplands of Northern Tablas covering areas within the bounds of three municipalities namely San Agustin, San Andres and Calatrava. Tiger grass (*Thysanolaena maxima*) or *luway* in the local dialect grows in abundance in these regions particularly in the villages of Doña Juana, Hinugusan, Binongaan, Mari-Sur, Victoria, Mari-Norte and Pagsangahan. Its flowers are popularly processed into softbrooms.

In 2010, a baseline study was collaborated by the Department of Trade and Industry and Romblon State University to profile the tiger grass industry in Mari-Norte. From the survey results, a need to extend the study to the six other adjoining villages was put forward. Likewise, the idea of coming up with a tiger grass resource map in Northern Tablas employing the prevailing mapping technology was deemed current and relevant.

This investigation was conducted to profile the tiger grass industry, the tiger farmers and soft broom processors in six adjoining tiger grass growing villages in Northern Tablas. Likewise, a comprehensive review of the biology and uses of tiger grass was compiled. Specifically the following questions were answered:

1. What do literatures report about the biology and uses of tiger grass?
2. What are the demographic profiles of the tiger grass farmers and /or processors in terms of sex, age, civil status and number of dependents?
3. What are the socio-economic profiles of the tiger grass farmers and/or processors in terms of educational attainment, estimated annual income, other sources of income and membership in organizations?
4. What are the profiles of the tiger grass farming industry in terms of the length of experience in engaging in the industry, distance of farm from home, estimated farm size, land tenure status and land use?
5. What are the farm inputs considering the farm expense, farm needs and human resource needs?
6. What are the farm outputs considering the annual production volume, price of produce, amount of produce sold and the estimated annual income generated from *luway* production?
7. What are the profiles of the tiger grass processing industry in terms of the length of engagement in the entrepreneurial activity, processing inputs, processing outputs like the products developed, pricing and the estimated annual income generated from soft broom making?
8. What are the practices employed by the processors in terms of product marketing and packaging?
9. What are the problems encountered by the respondents related to tiger grass farming and processing industries?
10. Where are the exact locations of the identified tiger grass farms including that in Marigondon Norte in San Andres, Romblon determined by the Global Positioning System (GPS)?

The respondents were distributed as follows: Victoria (7), Mari-Sur (29), Pagsangahan (30), Doña Juana (152), Binongaan (56) and Hinugusan (18). Of these 292 respondents, 55.8% were farmers, 40.8% were both farmers and processors, and 3.8% were solely processors. The industry profile data were collected through survey interviews using a validated questionnaire which was worded in Filipino. For the determination of the GPS coordinates, the survey team visited the actual tiger grass farms. Research data were processed and analyzed using SPSS version 11.5. Map data were plotted with the aid of a GIS software. Most of the data were disaggregated by village.

Results

1. There is a strong literature base and scientific evidence supporting the potential and significance of tiger grass, *Thysanolaena maxima* (Roxb.) O. Ktze. as a multi-purpose crop. Its economic importance is foremost advanced. This grass also holds a promise in phytoremediation, bioengineering, medicine, agriculture, climate change adaptation, and more.

2. There was a narrow gap in terms of percentage of males (58.2%) and females (41.8%) involved in tiger grass industry. Most of them were married (82.2%) with an average of four dependents. The mean age was 46 years old.

3. The socio-economic condition of the respondents was characterized by low educational attainment (elementary graduate or less) and low estimated annual income amounting to P24,000 only indicating that they belong to the social poor, if poverty is measured by these two indicators. Other income sources were limited to seasonal agricultural activities like copra production (43.5%) and palay farming (19.2%). There was a growing interest in social participation triggered by DSWD's 4Ps program (10%).

4. For the last five years, the increasing number of interested in tiger grass farming was observed. These farms averaging 1 hectare per farmer were commonly managed by the tenants or owned by the respondents. They were estimated to be 3 km away from their home. Under crops like coconut and palay were planted alongside tiger grasses.

5. The tiger grass farming industry required minimal farm inputs. Those with vast plantations spent around P5,000 on the average which they paid to farm workers who were usually relatives and family members. The common daily wage was P150 per day. Indigenous farm implements like bolo, tara- tara and tagad were commonly used in farm preparation and planting.

6. The annual production volume of tiger grass flowers ranged from 120 to 300 bundles per farmer wherein a bundle's cost ranged from P25 during peak months (harvest season, February to May) and P52 during off-peak months (October to December) yielding an average income estimate of P3,750 per farmer. However, a large percentage of the respondents refused to declare their income hesitant that they would be made to pay taxes or they would be removed from the list of DSWD's indigents. Should they cooperated in the survey, the income estimate could have been higher.

7. The average length of processors' engagement into soft broom industry was 10 years. Regular and jumbo soft brooms were sold from P25 to P60 respectively. The average capital needed in soft broom making was P3,000. This was self-funded. Soft broom making can assure every processor an annual income estimate of P17,500. But this result was again another understatement because of the misgivings of the respondents in revealing data related to their income.

8. Soft brooms had already established a number of markets within and outside the province but most of the products were concentrated in Doña Juana (53.5%). Marketing practices were of the traditional type but there were developments as to product packaging. Some processors were already aware of competition and were observing quality control, product branding (Tablas Premium) and product improvement. However, the use of 'Made in Baguio' labels were still prevalent.

9. The most common problems faced by the farmers were infestation of pests and animals (37.7%), theft (35.2%) and climate change (27.8%) while those of the processors were the tiresome processing activity (48.1%), lack of financial resource (25.6%) and mold attack during rainy days which were brought about by climatic changes (23.3%).

10. San Agustin (337.62-ha) has the largest plantation of tiger grass yet in the province as compared to those in San Andres (168.35-ha) and Calatrava (35.35-ha). Largest farm areas were in Doña Juana (215.52-ha), followed by Binongaan (88.50-ha), Hinugusan (33.60-ha), Pagsangahan (35.25-ha), Hinugusan (33.6-ha), Mari-Sur (30.25-ha) and Victoria (7.50-ha)., These values, along with the reported 130.6 hectares plantation in Mari-Norte show that there are about 541.22 hectares of tiger grass plantations altogether in Northern Tablas.

Recommendations

1. The identified resource maps in this study must be integrated into the Community-Based Resource Management map of the barangay, town and province for proper monitoring of the community's resources.

2. There is a need to fast track the formulation of a Tiger Grass R&D program that will set the roadmap towards consideration of the commodity as among the provincial, regional and national priorities. The RSU-Research, Extension and Production Unit can form a committee to organize the comprehensive R&D plan for tiger grasses particularly its industrial, medicinal, agricultural and environmental significance.

3. The Extension Unit of the College of Business and Accountancy of Romblon State University can initiate programs that will prepare these farmers and processors for a cooperative organization. Aside from pre-membership seminars, business –related knowledge on some aspects of the industry can also be extended. It can also package an information and education campaign on the potentials of tiger grass so that the community people will be made aware of its importance.

4. The developed prototype machine, the tiger grass pollen remover with woodworking tool, of the College of Engineering and Technology of Romblon State University can be tested in these barangays to unburden the farmers and processors of the physically taxing manual work. Likewise, a technology that can speed up the soft broom processing may also be explored upon.

5. The College of Business and Accountancy of Romblon State University in coordination with DTI, NEDA, DOST and DA can pilot the previously proposed "Entrepreneurial Camp" as an integration to the BSBA curriculum. This proposed encampment will be a two-month immersion activity of the business students in a potential community with economic activity to boast like tiger grass industry of these barangays. During the

encampment, the students will be taught and guided on the preparation of a project proposal and the best proposal could be packaged for possible funding. This activity is expected to broaden the social concept of the students and learn on hand what community development is all about.

6. On gender and development, a program can be packaged to empower women tiger grass farmers and processors by the GAD coordinators in the barangay or other agencies concerned.

7. On further studies, the following may be conducted or initiated:
   a. Profile of tiger grass industries in barangay Lusong and Salingsing in San Agustin;
   b. An *in situ* study of cultural practices on tiger grass farming focusing on plantation intervals and fertilizer applications;
   c. Cost analysis and market channel studies of soft brooms;
   d. Impact of climate change on the tiger grass industry; and
   e. Validation and trial studies on the reported uses of this plant as reported in literatures.

1 BACKGROUND OF THE STUDY

Tiger grass *(Thysanolaena maxima)* commonly called luway in the vernacular is a grass reaching a height of 2 to 3 meters. Its long and loose, diversely branching flower clusters at the ends of the stems consist of numerous fine slender branches having very fine delicate flowers and seeds. These are the materials used for soft broom production.

Tiger grass has been identified by Indian researchers as a perennial, high value, non-perishable cash crop for wide range of agro-climatic conditions. It is a multipurpose species which provides brooms, fuel, feedstuff and has high soil conservation value. The decoction of roots of this plant is used as mouth wash during fever. It has the comparative advantage of tolerance to harsh environmental conditions such as steep rocky mountain slopes, shallow soil, drought and high rainfall conditions. Therefore, it is suitable to grow on wastelands as well as in farms. The fibrous root system of the plant is very useful in checking soil erosion on steep slopes. After the harvest, the broom sticks (stem) are used as wall building material. The sticks have also been tried by paper and pulp industries for the manufacture of paper. The cultivation of this grass can wean away the practice of shifting cultivation and reduce the dependence of people on forests (Bisht and Ahlawat, 1998).

Luway grows in abundance in the mountainous regions of the three adjoining municipalities in Tablas island in the province of Romblon namely Calatrava, San Agustin and San Andres. The villages of Victoria, Doña Juan, Mari-Sur and Pagsangahan are located along an imaginary circle in the mountains of Central Tablas. Literatures reported that as of 2004, there were about 400 farmers in around 300-hectare tiger grass plantation all over the island (Servañez and Servañez). The industry has been around for decades but its potential remains unexplored. Records revealed that a number of interventions were made for the industry to gain market but they were short-lived. It was not until the Department of Trade and Industry (DTI) identified tiger grass as a crop for One Town, One Product (OTOP) program of San Andres and San Agustin that efforts were rekindled to help develop the potential of this industry (Fetalvero & Faminial, 2010).

In 2010, a baseline study was collaborated by DTI and Romblon State University (RSU) to profile the tiger grass industry in Marigondon Norte in San Andres. The project intended to cover other areas in San Andres like Mari-Sur, Victoria and Jun Carlo. However, due to limited budget and time constraints, the researchers limited the study's scope in Marigondon Norte where 100 farmers were interviewed. These initial data were used by DTI in planning developmental and strategic interventions in promoting the tiger grass industry of Tablas island in the region and beyond.

Similar assistance was provided by the Southern Tagalog Agricultural Resources Research and Development Consortium (STARRDEC). In August of 2010, experts from STARRDEC recommended that tiger grass be prioritized as a commodity in the province. However, its general production volume must be profiled first before this could be considered in the Regional Development Council (RDC). This is the reason why RSU is advocating tiger grass as a potential commodity for research and development.

Statement of the Problem

This study was conducted to profile the tiger grass industry, the tiger farmers and soft broom processors in six adjoining tiger grass growing villages in Northern Tablas namely Binonga-an, Hinugusan, Doña Juana, Victoria, Mari-Sur and Pagsangahan in the province of Romblon. Likewise, a comprehensive review of the biology and uses of tiger grass was compiled. Specifically the following questions were answered:

11. What do literatures report about the biology and uses of tiger grass?
12. What are the demographic profiles of the tiger grass farmers and /or processors in terms of sex, age, civil status and number of dependents?
13. What are the socio-economic profiles of the tiger grass farmers and/or processors in terms of educational attainment, estimated annual income, other sources of income and membership in organizations?
14. What are the profiles of the tiger grass farming industry in terms of the length of experience in engaging in the industry, distance of farm from home, estimated farm size, land tenure status and land use?
15. What are the farm inputs considering the farm expense, farm needs and human resource needs?
16. What are the farm outputs considering the annual production volume, price of produce, amount of produce sold and the estimated annual income generated from *luway* production?
17. What are the profiles of the tiger grass processing industry in terms of the length of engagement in the entrepreneurial activity, processing inputs, processing outputs like the products developed, pricing and the estimated annual income generated from soft broom making?
18. What are the practices employed by the processors in terms of product marketing and packaging?
19. What are the problems encountered by the respondents related to tiger grass farming and processing industries?
20. Where are the exact locations of the identified tiger grass farms including that in Marigondon Norte in San Andres, Romblon determined by the Global Positioning System (GPS)?

Significance of the Study

The success stories of small and medium enterprises that were able to penetrate the market through the Department of Trade and Industry's One Town, One Product Program (OTOP) inspired other towns with rich resources to do the same. Baseline data from this study are important for government agencies like Department of Science and Technology, Department of Trade and Industry, and Department of Labor and Employment for any possible program or project they could introduce in the area. The local government units of San Andres, Calatrava and San Agustin particularly the barangay councils of Mari-Sur, Victoria, Pagsangahan, Binongaan, Hinugusan and Doña Juana will also benefit from this study.

Funds are appropriated by the DOST, DTI, DOLE and other government agencies for income-generating projects like tiger grass industry. In order for these funds to be downloaded, project proposals are required. The rich data gathered in this study will help project proponents to plan out better project proposals from tiger grass farming to tiger grass processing. Through the results of this study, decisions as to where the starting point will be in helping the industry can be better made.

The local government of San Andres, Calatrava and San Agustin will be able to update their community resource map through the data from this study. These data could also serve as benchmarks in initiating policies about the industry or in introducing interventions to improve the tiger grass products, the farmers' behavior and their farming and marketing practices.

The Research and Development Unit of the Romblon State University can review the baseline data generated by this study and examine in which aspect it can be of help in developing the tiger grass industry in these adjoining villages. The unit can take the lead in tiger grass knowledge and technology transfer and techno-demo farms establishment.

Identification of the exact location of the tiger grass farms can help upgrade the community resource map of the town and could facilitate immediate monitoring when disasters affect the province.

Scope, Delimitations and Limitations of the Study

This study was conducted to profile the tiger grass industry, the tiger farmers and soft broom processors in six adjoining villages in Northern Tablas in the province of Romblon. Villages included in the study were Doña Juana, Binongaan and Hinugusan in San Agustin, Mari-Sur and Victoria in San Andres and Pagsangahan in Calatrava. Plantations in Mari-Norte were included in GIS mapping. Based on DOST's claim that about 400 farmers (Servañez & Servañez, n.d.) are engaged in the tiger grass industry, complete enumeration technique was employed. Existing databases were used to locate the farmers and/or processors. Key informants were also interviewed. One limitation of the study was the refusal of some respondents to share data particularly those that were related with their income from farming and processing with the apprehension that they would be taxed or removed from the list of indigent families.

2 THE BIOLOGY OF TIGER GRASS

Thysanolaena maxima (Roxb.) O. Ktze[1]. belongs to the *Poaceae* or *Graminae* family. It is commonly called *tiger grass*, *broom grass* and *bouquet grass* in English; *thong kong, lao laeng, yaa mai kuat* and *yaa karb phai* in Thai; *dot, dong trung hoa thao, cay le, ong anh, say* in Vietnamese; *khem khong, yaa yung, dok khein, bong kha ching* and *tau khaou thuwa* in Laos; *jhadughas* in Hindi; *amliso* in Nepalese; *taza* in Nishi; *kamgang* in Adi; *eppane-nani* in Apatani; *phool jhadu* in Assamese (Bisht and Ahlawat, 1998); *chorondora* in Bangladesh (Khisa et al., 2001) and *ons/kuchi* in Uttarakhand, India (Bhuchar, 2008). In the Philippines, it is locally called *boi-boi* in Ilocano, *tambo* in Tagalog (Baldino, 2002), *lasa* in Bicol, *sugbo* in Catanduanes and *luway* in Romblon (Fetalvero et al., 2011).

T. maxima has been reported to be growing most in Southeast Asia. Literatures reported its prevalence in Bangladesh (Pal, 1991; and Palni et al., 1994 in Bhuchar, 2008); Bhutan (Palni et al., 1994 in Bhuchar, 2008); Cambodia; Darjeeling Himalaya (Shankar et al., 2001); Nicobar Island (Saikia et al., 1992); Megahalaya (Lyngdon and Baishya, 2010) in India (Saikia et al., 1992; Ramm Botanicals, 2009; Weed Watch Australia, 2011; Palni et al., 1994 in Bhuchar, 2008); Indian Ocean Islands; Indonesia; Laos (Vernon, 2006 in Nicholson et al., 2008); Malaysia (Palni et al., 1994 in Bhuchar, 2008); Myanmar (Maki et al. 2007; Palni et al., 1994 in Bhuchar, 2008); New Guinea (Palni et al., 1994 in Bhuchar, 2008); Tehrathum in Nepal (Palni et al., 1994 in Bhuchar, 2008); Sri Lanka; Thailand; Nam Dong (Wetterwald et al., 2004) in Vietnam; Japan (Weed Watch Australia, 2011); and in the Chinese provinces (Palni et al., 1994 in Bhuchar, 2008; Weed Watch Australia, 2011) of Xishuangbanna (Tang et al., 2006), Guangdong, Guangxi, Guizhou, Hainan, Taiwan, and Yunnan. In the Philippines, tiger grass is widely distributed in Ilocos Norte, Apayao, Bontoc, Benguet (Baldino, 2002), Nueva Viscaya, Carranglan in Nueva Ecija (Tuddao and Evasco, 1997), La Union (Abad, 2008), Bulacan, Zambales, Bataan, Laguna, Tayabas, Sorsogon, Milangnilla (Quiachon, 2002) in Cebu, Mindoro, Palawan, Batangas (Caringal and Bañados, 2008), Romblon (Fetalvero et al. 2011) and other places in Mindanao.

T. maxima is a tall reed-like tufted (Saikia et al., 1992) non-invasive and very vigorous perennial grass (Ramm Botanicals, 2009) with leaves bearing resemblance to those of bamboo (Bhuchar, 2008). It is a non-timber species (Baldino, 2002) forming a large and dense clumps (Ramm Botanicals, 2009) producing numerous upright or arching stems (up to 10 mm thick) that are unbranched and have joints at regular intervals. These stems bear large alternatively arranged leaves. Culms are solid, smooth and rounded and grow up to a height of 4 m (Saikia et al., 1992). The long lance shaped leaves (Ramm Botanicals, 2009) are relatively broad (25-60 cm long and 3-7 cm wide) with pointed tips and entire margins. They consist of a sheath at the base, which encloses the stem, and a spreading leaf blade (Weed Watch Australia, 2011). The inflorescence that is about 30 to 90 cm long resembles a foxtail (Bisht and Ahlawat, 1998). It thrives in low to medium elevations but grows faster in higher elevations (Noble, 1991 in Baldino, 2002).

Technically Bisht and Ahlawat (1998) described *T. maxima* as a huge tufted grass, up to 3 m tall, culms solid, leaf-sheaths at least the upper ones, tight, glabrous, terete, smooth, the nodes glabrous, margins with some short stiff hairs towards the throat; blades lanceolate-acuminate, abruptly contracted to a short petiole for a subcordate base,

[1] *Thysanolaena maxima* (Roxb.) O. Ktze., Rev. Gen. Pl. 2: 794. 1891; Hitchc.in Lingn. Sci. Journ. 7: 207. 1931, Man. Grass. U. S. 569.1951;Ohwi in Acta Phytotax. Geobot. 10: 272. 1941; Keng, Fl. Ill. Pl. Prom. Sinicarum Gram. 340. f. 280. 1959; Bor, Grass. India 650. 1960; Hsu in Hara, Fl. E. Himalaya 378. 1966, in Taiwania 16: 227. 1971, Taiwan Grass. 237. pl. 13. 1975; Backer & van den Brink, Fl. Java 3: 528. 1968; Gilliland, Grass. Malay 55. F.5.1971.

acuminate to a fine point, glabrous, the margins scaberulous, up to 50 cm long and 7 cm wide; ligule a shallow membrane 1-2 mm deep, backed by short stiff hairs; Inflorescence a huge and drooping panicle 60 - 90 cm long or more wide at anthesis, the axis and branches at first rounded, ultimately, capillary, not sharply angled; spikelets numerous, often in pairs on a common peduncle, each pedicel distinct; lower glume clasping, ovate-acute, obscurely 1 nerved, up to 6.5 mm long; upper glume more transparent; lowerlemma lanceolate-acuminate, sub-hyaline, with 1 or 2 long setose hairs near the margi; upper lemma lanceolate-acuminate, 3 nerved, green between the nerves, hyaline thence to the margin, with stiff setose hairs along the hyaline portion on both sides; palea a narrow, 2 nerved, hyaline scale; stamens 2 (3); stigmata 2, plumose; reddish brown, the rachilla continues as a flattened process with an expanded tip, beyond and behind the upper lemma. The aspect of the spikelets changes with the onset of anthesis when the upper lemma emerges and its setose hairs gradually adopt a stance at right angles to the lemma's surface

T. maxima grows well in temperate and sub-tropical regions in wide ranging habitats like hillsides and valleys, among rocks, in thickets, forest margins, open grasslands, river banks and can be successfully cultivated in margins of rainfed and irrigated agriculture fields, degraded and wastelands, forests and along roads, footpaths (Bhuchar, 2008), on marginal lands, wastelands and jhum fallow on a wide range of soils varying from sandy loam to clay loam. The planting can be done by seeds, rhizomes (Bisht and Ahlawat, 1998) or slips – containing 3-4 buds and 1-2 nodes (Bhuchar, 2002) mixed with other crops in open fields (Nicholson et al., 2008).Under natural condition, it regenerates through seeds (Bisht and Ahlawat, 1998). It has a stable seed input and even distribution (Tang et al., 2006). The flower spikelets appear hairless at first, but as they mature the small hairs become more obvious, giving the seed-head a slightly feathery appearance. These spikelets contain a small and feathery seed (about 0.5 mm long) which can easily be spread about by wind, water, vehicles and mowing equipment (Weed Watch Australia, 2011). When the seeds mature usually from February to March, the numerous tiny flowering spikelets (1.5-2 mm long) are shed from the large seed-heads (30-60 cm long). The seed germinates best in the beginning of the rainy season on loose and exposed areas (Bisht and Ahlawat, 1998) under light conditions at 25°C. Nursery can be established during warm summer months and a reproducible micro-propagation protocol with the help of excised zygotic embryos for mass multiplication of *T. maxima* has already been developed for this purpose (Bhuchar, 2002).

T. maxima plantation has a cycle of about six years (Shankar et al., 2001) but could reach a life span of 10 years or more when properly managed (Baldino, 2002). The number and length of its culms increase up to the third harvest and decline thereafter. Domestication is likely to be facilitated if the species is adaptable, market demand is greater than the production in natural populations, profitability from cultivation is high, and there are not many job opportunities or sufficient agricultural landholding with the forest dwellers (Shankar et al., 2001). It grows in tussocks and the culms arise centrifugally during the peak growing season and bear inflorescence (panicle) on shoot apex at the end of vegetative growth (Bisht and Ahlawat, 1998). The tassel-like flowers form at the stem ends during summer (Ramm Botanicals, 2009). Collection of tiger grass flowers for soft broom making can start after 1 -2 years of plantation (Nicholson et al., 2008). The number of flowering shoots had high significant positive association with plant height, tillers per plant, rachis per plant and panicle length both at genotypic and phenotypic levels (Jagadev and Patnaik, 1994). When the panicles are cut, the stem portion (3–4 m) is left out in the field and is burnt (Saikia et al., 1992).

The literature on this species is scanty and no scientific study had been made so far on intraspecific variations, genetic improvement, developing cultivars and cultivation practice. However, work on genetic improvement and development of suitable cultivation technique has already been started at the State Forest Research Institute in Arunachal Pradesh, India . The parameters being identified for improvement are sprouting ability of clones or number of culms (tillers) per clump, length of panicle or brooms, height and growth of clumps, toughness of rachis and softness of ultimate floral branches, number of floral branches per inflorescence, and disease and pest resistance. Considerable variations have been observed in color and size of the inflorescence and number of culms per tussock, which are the main economic part of the plant. Some plants were found to have strong, brown and short inflorescence and fetch more price to the growers while the other ones with long, green, and slender inflorescence fetch less price in the market. Desirable morphotypes have been multiplied and are being distributed among farmers (Bisht and Ahlawat, 1998).

T. maxima has a high benefit cost ratio and very good market, processing and value addition facilities. As a result, its cultivation is expanding rapidly. Even without any external intervention, the farmers are getting good returns. In villages where the farmers have taken up the cultivation of this crop, within 10-15 years it has almost completely occupied all the lands previously used for shifting cultivation. It has a low starting cost and quick returns (Tiwaki, 2001).

Propagation, Plantation Development and Management
(DENR-CAR, 1992)

In the uplands like in the Cordilleras, the following propagation, plantation development and management technologies are being recommended by DENR-CAR.

Techniques in propagation. There are two known methods of propagating tiger grass namely, by seeds and by rootstocks. However, propagating by seeds is rarely done because spacing between plants cannot be regulated and will take a long time for the plant to mature and produce panicles. Because of this, propagating tiger grass by rootstock is commonly used. Clumps of mature tiger grass are uprooted, and the upper portion of the grass is cut. About 12-15 cm of the culm measured from the rootstock is left and can be divided into 3 culm individual rootstock.

Site preparation. Rootstocks of tiger grass can be planted directly in prepared planting sites. But for better results, the following should be observed: Planting sites should be prepared by completely removing the grass and other undesirable vegetation; strip clearing or spot-ring clearing methods may be employed; and planting holes of about 30-50 cm in diameter shall be prepared for the rootstocks.

Method of planting. As mentioned earlier, Tiger grass can be propagated by seeds or by rootstocks. But for rootstocks, they are planted in prepared planting holes. Planting holes of about 30-50 cm in diameter are prepared first before the rootstocks which are collected from the mother plants are finally planted.

Planting season. For best results, the rootstocks should be planted during the onset of the rainy season. The plants will produce more shoots and could yield more panicles.

Spacing between plants. The ideal spacing requirements of Tiger grass when planted in pure plantation is 4 m x 4 m. If interplanted with other crops like fruit trees, the recommended spacing is 8 m x 8 m.

Soil and fertilizer requirements. Tiger grass can thrive in many types of soil such as clay, sandy clay loam, and sandy loam. However, fertilization can be done in areas where soil nutrients are deficient to enhance growth and flower development of the plants. The appropriate fertilizer application is 20 grams of NPK (or complete fertilizer) per rootstock or plant. Fertilizer should be applied 10 cm away from the base of the plant and dug around to mix the fertilizer with the soil.

Some Cultural Management of the Plantation. The success of a tiger grass plantation is dependent on the cultural management schemes to be employed. The following are the recommended management practices to be undertaken:

- One year after planting, inventory of the mortality should be conducted so that replanting can be immediately done during the next rainy season.
- To minimize competition of soil nutrients, space, light and moisture, weeding should be done during the rainy season when there is abundant growth of competing vegetation.
- Fertilization may be done in areas deficient of soil nutrients to enhance growth and flower development following the rate mentioned earlier.
- Plantations should be protected from stray animals and wild fires.

Harvesting of Panicles. When the panicles reach a length of about 70 cm and more, they can already be harvested. The following are some recommended techniques when harvesting Tiger grass:

- Cut the panicles with a sharp sickle or bolo when they are still green and soft. This could ease the cleaning and removal of seeds. Mature ones are hard and rough, and could result to low quality brooms.
- When the panicles have been harvested, cut about 90 percent of the aerial part of the plant so that more tillers/shoots will develop during the next rainy season.
- Harvested panicles must be dried under direct sunlight for about 2-3 days.

- During and after the drying period, the seeds must be removed by shaking or patting them lightly against a big stone or a concrete pavement.
- After drying and cleaning, panicles must be bundled into an average size of about 7 cm in diameter per bundle. Each bundle consists of an average of 210 panicles.
- Bundled panicles are then ready for softbroom-making and marketing. The panicles are then sold and delivered to softbroom manufacturers.

Season of Harvesting. Panicles usually develop from October to March. The best time therefore to harvest the panicles is during the months of February and March. Make sure that the panicles reaches the desired length, softness and greenness to produce quality softbrooms. The age of the panicles to be harvested is 5 months.

3 USES OF TIGER GRASS

This section details the evidences on the claim that the grass is a multi-purpose crop. Literatures reported manifold uses of tiger grass in industry, medicine, agriculture, ecology and environmental management, among others.

Industry

Brooms

T. maxima is a perennial, high value, non-perishable cash crop and a multipurpose plant (Bisht and Ahlawat, 1998; Rai and Sharma, 1994 in Bhuchar, 2008). Its inflorescences are widely used for the manufacture of soft brooms (Saikia et al., 1992; Rai and Sharma, 1994 in Bhuchar, 2008; Bisht and Ahlawat, 1998; Ronya, 1998; Shankar et al., 2001; Baldino, 2002; Fu et al., 2003; Saigal & Sharmistha, 2003; Wetterwald et al., 2004; Pandit & Thapa, 2004 in Dogan et al., 2008; Bhuchar, 2008; Dogan et al., 2008; Jain, 1981 in Bhardwarj and Gakhar, 2008; Nicholson et al., 2008; Aryal et al., 2009; Ramm Botanicals, 2009; Lyngdon and Baishya, 2010; Fetalvero, 2011; Marier, 2011; Namsa et al., 2011) which are good sources of income generation (Malla & Chhetri, 2009). *T. maxima* brooms are preferred for sensitive surfaces (Resurrecion, 2000 and Sharma et al., 2001 in Dogan et al., 2008) like wooden, stone and mosaic floors, and carpets and is superior to some of the other broom species like *Cocos nucifera, Phragmites* spp. and *Saccharum* spp (Bisht and Ahlawat, 1998; Bhuchar, 2008).

An eight-month old tiger grass which starts producing panicles could immediately generate cash when harvested and sold as raw materials to broom manufacturers. These are directly made into soft brooms (Baldino, 2002). *T. maxima* flowers are usually available in large quantities from July to February (Bisht and Ahlawat, 1998).

T. maxima brooms processed from inflorescence bearing panicles have a high demand in countries like India, Nepal, Bangladesh and in the Middle East. Shankar et al. (2001) estimated a return of 1.7 times of the total investment during a six year plantation cycle, which included the cost of labor (62%) of the input and rent for the land (about 33%). The returns could increase significantly if the cultivators are self-employed and the land is available free of rent. The same study estimated that each ton of tiger grass flowers after processing into finished brooms fetched about US$1,333 (Bhuchar, 2008).

A reliable estimate of the annual broom demand in different countries is not available (Bhuchar, 2008). In India, tiger grass cultivation has already been successfully taken up as commercial plantations in the states of Meghalaya, Sikkim, West Bengal, etc. (Ronya, 1998). In the district of Arunachal Pradesh, it has a recorded high annual market returns and remains always in high demand in local, regional and national market levels (Sarmah and Arunachalam, 2011). The annual Indian broom market was estimated to be about US$60 million (Shankar et al. 2001 in Bhuchar, 2008). In Chepang, Nepal, the market demand for soft brooms was reported to have significantly increased (Aryal et al., 2009). Judging from the current trend of its utilization, the demand is steadily growing and marketing should not be a problem (Ronya, 1998).

In Laos, tiger grass is the country's second largest NTFP export (200 tons a year in 2006). There is a plentiful supply of natural resources to supply market demand. One hectare can yield a gross income of US$503 (Vernon, 2006 in Nicholson et al., 2008). Brooms made in Laos, whose raw material almost doubled between 1996 and 1999 are sold in the domestic market. There are potential new markets in neighboring countries such as Vietnam, China and Japan. The cost benefit analysis shows more benefit for farmers to trade in finished brooms instead of selling raw materials (Nicholson et al., 2008). A wiki page (Dok khaem) reported a unique and new species of tiger grass in Laos yet to be verified that distinguishes between its male and female flowers. It says:

> "A new variety of tiger grass, introduced to Houn district in Oudomxay by a Thai investor, has more bristles and produces different 'male' and 'female' stems. The female stems are preferred for broom making, as they have many more bristles. The taxonomy of this new variety is not yet clear."

In Southern Batangas in the Philippines, sales from non-timber forest product like tiger grass are means for much needed cash for household essentials and other necessary commodities (Caringal and Bañados, 2008). In Romblon, about 50 percent of the tiger farmers' annual income came from the production of *T. maxima.* If its 2009 production volume is sustained, an estimated annual revenue of US$23,000 to US$46,000 can be realized, much higher if these are processed into soft brooms (US$84,000 to US$170,000) (Fetalvero et al., 2011).

Pulp and paper

The culms of *T. maxima* have been tried for the manufacture of paper (Bisht and Ahlawat, 1998; Rai and Sharma, 1994 in Bhuchar, 2008). Fibers, averaging 1.25 mm in length at 45% yield (unbleached), could be obtained from this grass. The laboratory handmade paper sheets from *T. maxima* exhibited good properties, with a burst factor of 30, a breaking length of 3,555 m and a tear factor of 106. Hence, it can be suggested that *T. maxima* can become a potential source of raw material for pulp- and paper-making either alone or in combination with the conventional pulp- and paper-making raw materials. This material could help meet the future demand for pulp- and paper-making raw material, if properly exploited (Saikia et al., 1992).

Its pulps were also processed into insulation boards. The boards had very good strength but moderate heat insulating properties. In terms of moisture resistance properties, they compared favorably with the imported ones (Razzaque and Khan, 1978). It has been reported that *T. maxima* leaves were also tried as substrates for cellulase and ethanol production (Yimyong et al., 2005).

Medicine

Etnomedicinal and ethnobotanical studies in India and Nepal reported significant uses of *T. maxima* in traditional health care. Folks of North Sikkim in India reported that a decoction of 200-300 grams of its young roots for one dose is used twice a day to treat bronchial problem. Poultice of young flowers is used in rheumatic pain and skin swelling (Maity et al., 2004). The roots are used in flatulence (Jana and Chauhan, 2000). Lodha people prescribe a paste of the flowers along with country liquor and honey as contraceptive to women. A paste of its dried or fresh roots is applied on the skin to check boils (Rai, 2003) while the decoction of roots is used as a mouthwash during fever (Bisht and Ahlawat, 1998; Rai, 2003), treatment of halitosis (Singh et al., 2002) and along with common salt is used as remedy for mouth sore (Pal and Jain, 1998).

In Meghalaya, traditional healers and village folks make a paste from its inflorescence mixed with a pinch of slaked lime as herbal remedy for the treatment of boils (Singh et al., 2002, Hynniewta and Kumar, 2008, and Malla and Chhetri, 2009) and cancer (Hynniewta and Kumar, 2008). Young stem juice is applied on the eye when eyes become red and dirty (Kharkongor and Joseph, 1981 in Hynniewta and Kumar, 2008).

As an anthelmintic, two teaspoonfuls of root juice (5 pieces) about 10cm long, crushed by mortar and pestle are given twice a day for 2-3 days to patient (Shrestha, 1985; Mahato and Chaudhary, 2005). A scientific study about its anti-microbial property found out that its roots could positively respond against bacterial strains like *Pseudomonas aeruginosa, Staphylococcus aureus, Bacillus subtilis and Escherichia coli* (Mahato and Chaudhary, 2005).

Agriculture

Several studies reported that the tender culms, leaves and tips of *T. maxima* are used as fodders for cattle during lean period (Saikia et al., 1992); Bisht and Ahlawat, 1998; Shankar et al., 2001; Baldino, 2002; Dogan et al., 2008; Rai and Sharma, 1994 in Bhuchar, 2008; Dogan et al., 2008; Malla and Chhetri, 2009; Roothaert et al., n.d.). In Myanmar, captive elephants in the Okkan Reserved Forest fed on *T. maxima* (Himmelsbach, 2006). Rumen degradability of *T. maxima* after 48 hours was observed to range within 404 to 488 g/kg (Huque, et al., 2001). However, Kafle (2005) claimed that although this grass species produces more foliage, its palatability is less. It also causes haematuria among cattle and buffaloes that fed on them Joshil and Singh (1989).

In Nepal, local knowledge of farmers was used as basis for planning ruminant diets. Farmers ranked *T. maxima* sixth out of the 14 preferred fodders for milk however there was a negative animal response (-7%) (12th). Animal response is the difference between average production during experimental and pre-experimental feeding. For butterfat production, *T. maxima* was the second most preferred fodders by the farmers and showed a 6% animal response (9th) (Subba et al., 2002). *T. maxima* was also ranked third as farmers' preferred feeds for livestock and was reported to stimulate milk production and observed to be palatable to animals even during rain or cold weather conditions. However, because of its lower leaf to stem ratio, it produces less fodder (Livestalk, 2011).

In a study conducted by Subba et al. (2004), *T. maxima* was investigated in terms of appetite satisfaction among 8 buffaloes. Duration of appetite satisfaction was defined as the time between refusal and occurrence of behavior indicative of hunger. This grass satisfied appetite of buffalo for 380 mins.

Rohilla and Bujarbaruah (2000) reported that tiger grass can be fed to rabbits and will improve performance if it is processed (dried and ground) and mixed with concentrate mixture up to 40% level. In their experiment, 18 rabbits (10-12 weeks old) were divided into three groups. The first group was fed solely on fresh tiger grass leaves (T1), the second group was offered 100% dried and ground tiger grass (T2), and a combination of 40% dried and ground tiger grass with 60% concentrate was given to the last group (T3) for a period of 105 days. Tiger grass feeding had a significant ($p \leq 0.05$) effect on growth and dry matter intake (DMI) of rabbits. Daily weight gain of rabbits averaged 9.76, 11.78 and 15.73 g/day respectively for groups T1, T2 and T3. Average daily DMI was recorded as 106.99, 112.65 and 115.72 g/day in corresponding treatment groups.

Based on nutritional analyses, *T. maxima* leaves are found to contain a balanced proportion of nutrients making them as good forage and fodder for livestocks (Table 1).

Table 1. Nutritional analysis of *T. maxima* leaves (Bhuchar, 2008)

Parameter (%)	Palni et al. (1994)	Singh et al. (1995)	Bhuchar (2002)
Digestibility	57.9	-	54.3 – 57.9
Total Ash	11.8	5.65	10.7 – 11.8
Ether extract	6.67	1.94	4.2 – 6.7
N-free extract	33.1	51.6	39.3 – 44.6
Crude protein	18.1	10.2	15.1 – 18.2
Crude fiber	30.4	30.5	29.5 – 31.0
Cellulose	30.2	-	30.3 – 37.8
Hemicellulose	29.6	-	29.6 – 34.4
Lignin	9.1	-	4.6 – 9.2

Ecology

Soil and Water Conservation and Bioengineering Device

T. maxima has also been reported as a good soil and water conservation species. Its cultivation promotes the sustainable use of fragile and degraded lands (Shankar et al., 2001; Bisht and Ahlawat, 1998). The roots of the plants which are used in windbreaks or shelter belts bind the soil and protect topsoil and nutrients from erosion on sloping terrain, agricultural fields (Bisht and Ahlawat, 1998; Nicholson et al., 2008; ECS, 2008) and landslide (Mathema and Joshi, 2010). It is also used as a backup fodder grass on contour strips and terrace risers (Stapleton, 1989), a good soil cover, a crop to maximize land use, a tool for the management of pine forest, a protection from forest fires (Baldino, 2002) and a slope stabilizer in the hills and mountains. These techniques are nature-friendly, cost-effective and sustainable (Sharma, 2004).

In a three-year investigation in the Philippines, contour planting with tiger grass as a biological measure to reduce soil erosion in newly burned pine watershed showed better performance in terms of vegetative cover, surface runoff and erosion yield than the control and was equally comparable to broadcast sowing and contour trenching. It was the cheapest among the treatments investigated for revegetation and rehabilitation (Costales, 1985).

CHIAT or Contour-Hedgerow Intercropping Agroforestry Technology is a farming technology used in reducing soil erosion and improving soil fertility (Partpa and Watson, 1994 and Palmer, 1996 in Khisa, 2001). In Bangladesh, *T. maxima* was one of the two grass species that came out successful of the 22 NTFPs tested for suitability as hedgerow species in terms of runoff and soil erosion control. Landslides have been effectively controlled by hedgerows with *T. maxima.* Monitoring test shows that contour hedgerow reduced soil loss by 55-80% and runoff by 30-70% (Khisa, 2001) after three years of investigation. Following the series of tests in Bangladesh, Khisa (2001) characterized this hedgerow species as either seed or rhizome propagated, fast-growing, tolerant to pruning, high biomass production and drought and fire tolerant. The grass is also used as erosion control, green manure, fodder, firewood and cash crop.

The plantation of *T. maxima* was successfully done along with *Cajanas cajan* in degraded shifting cultivation. Soil samples from the experimental plots were collected at different stages and analyzed. Soil fertility and productivity were found to be better when *Thysanolaena maxima* was planted with *Cajanas cajan* than as a sole crop (RFRI, 2007). Cultivation of tiger grass on agriculture terrace margins improves forage production and soil conservation without affecting the productivity of the crops. It was reported that 1 kg of fresh rhizome planted at a distance of 1 x 1 m yielded about 16 ton/ha of above ground biomass in the third year of plantation (Bhuchar, 2008).

However, performance of erosion control measures is location specific. Hence, effect of stone bunds and vegetative barriers like *T. maxima* on erosion, crop yield and soil properties in degraded hill slopes was studied for three consecutive years (2000–2002) in Eastern India. Simultaneously, the morphological parameters of vegetative barriers were also evaluated. Results revealed that tiger grass was the tallest plant among all barriers. Although it was not as comparable with the other two grasses studied (*Vetiveria zizanioides* and *Saccharum* spp.) it was found to be an effective vegetative barrier in controlling soil erosion, improving crop yield and restoring soil fertility (Sudhishri et al., 2008).

Whenever possible, broom grass is used in bioengineering as an effective and low cost measure. Sharma et al. (2001) and Mathema and Singh (2003) in Bhuchar (2008) found out that plots treated with broom grass can reduce water runoff and soil loss by up to 88% as compared to bare land.

Table 2. Water and soil runoff quantities and conservation values of different land use types in Khanikhola watershed in Sikkim, India (Sharma et al., 2001 in Bhuchar, 2008)

Plot type	**Water runoff (liter)**	**Soil loss (kg)**	**CV water (%)**	**CV soil (%)**
Maize	25±4	1.51±0.18	69	56
Finger-mullet	18±3	1.32±0.14	78	62
Mixed cropping	12±3	0.95±0.12	85	73
Large cardamom	15±3	0.45±0.06	81	87
Broom grass	10±2	0.41±0.07	88	88
Bare land	80±11	3.46±0.35	-	-

CV: Conservation value

Table 3. Comparison of soil loss and runoff from different land use types in Nepal (Mathema & Singh, 2003 in Bhuchar, 2008).

Land use type	Soil loss (t/yr/ha)	Runoff (%)
Outward sloping terrace	10.4	2.8
Degraded land	21.3	40.1
Degraded land treated with broom grass	12.8	16.5

According to Clark and Hellin (1996) in Kafle (2005), engineering functions of grasses can be categorized into six: catch, armor, reinforce, anchor, support and drain. Armor, not anchorage is the main function of shallow rooted species while catch, reinforce and drain are their secondary functions. Kafle (2005) evaluated the performance of *T.* maxima along with other four grasses as a bioengineering device. *T. maxima* was found to have an excellent catch, moderately useful armor, excellent reinforce and moderately useful support. It was the most effective grass in reinforcing the soil by providing a network of strong roots that increases the soil's resistance to shear. It moderately supports the soil mass by its strong and long fibrous roots and has excellent interception, storage, leaf drip because of its great height (up to 4.9 m), foliage lateral spread (up to 5.14 m) and large clumping characteristics. In terms of its hydrological functions, it has excellent soil binding capacity and ground surface protection, interception, storage, leaf drip but moderate infiltration due to its strong fibrous roots catching the soil firmly leaving few pores in the soil.

Kafle (2005) further reported that the effective spacing of *T. maxima* for bioengineering purposes is 2.4 m in plain and 1.8 m in slope with maximum effective rooting depth of 0.5 to 1 meter (Howell, 1999 in Kafle 2005). Its height ranges from 3.24 m to 4.91 meters and foliage spreads laterally from 3.87 m to 5.13 m. Its root extends vertically up t 9.5 m and spreads laterally up to 1.32 m horizontal. Its foliage protects the ground surface area from direct raindrop effect up to 82.87 m^2. Its root binds up to 5.19 m^3 soil (Kafle, 2005).

Table 4. Characteristics of Tiger Grass as a soil bioengineering device (Kafle, 2005)

Thysanolaena maxima	
Character Large clumping **Distribution** Terai to 2,000 m **Sites** Varied **Uses** Mainly used in bioengineering, brooms, fodders **Patterns of Retaining Molasses** Terrace boundary, marginal land and degraded land, stream bank **Insect/pest/disease/rodents condition** No insect/disease problem **Rooting depth** 70 to 95 m **Root lateral spread** 103 to 132 cm radial **Height** 3.2 to 4.9 m **Ground surface area protected by foliage against direct raindrop effect** 471,920 to 828,715 m^2 Mean: 664,954 m^2	**Volume of Soil bound by roots** 2,339,529 to 5,197,475 cm^3 Mean: 3,778,119 cm^3 **Spacing** Plain: 240 cm Slope: 180 cm **Quantity of planting material required** *For single row* of 5m, 10m, 25m, 50m, 75m, 100m length, quantity of planting material is 4, 10, 21, 31 and 41 respectively. *For double row* of 5m, 10m, 25m, 50m, 75m, 100m length, quantity of planting material is 5, 11, 29, 62, 92 and 122 respectively. **Shade effect** Max. 6.8m for mean height 3.9m **Type of root** Fibrous **Propagation** Slip cuttings

Phytoremediation

Literatures also reported about the phytoremediation potential of *T. maxima* in wastewater treatment and stabilization of mined out areas. Tiger grass can be planted in degraded soils where other plants do not grow (Nicholson et al., 2008). It is currently being explored in the recultivation experiments on horizontal areas and on slope areas at the open pit mine Nui Beo in Vietnam. According to the first monitoring results, tiger grass species developed well and much faster than the tree species (Broemme and Stolpe, 2011). In Guangxi, China the ability of *T.*

maxima to accumulate antimony (Sb), a toxic element and a global environmental contaminant is also being tried out (Zhang, 2009).

The findings of the experiment of Rotkittikhun et al. (2007) revealed that *T. maxima* is comparable to vetiver grass, *Vetiveria zizanioides,* for phytostabilization of Bo Ngam lead mine in Thailand. It shows very high tolerance to lead concentrations in its roots and shoots. The application of inorganic fertilizer (150 mg/kg) improved its growth and its uptake of Pb. However, when the soil is amended with pig manure, it cannot tolerate high electrical conductivity thereby reducing the amount of uptake by roots and transport of Pb to shoots. Rotkittikhun et al. (2007) specified that *T. maxima* can withstand lead concentrations up to 100,000 mg/kg.

T. maxima can also be a suitable species for wastewater treatment systems. Under experimental conditions, the response of *T. maxima* in its early stages of growth to different levels of phosphorous supply and varying water depths was investigated. Nitrogen and phosphorous concentrations in the water sediment and different plant parts were studied. Nitrogen concentration in the leaves of *T. maxima* increased three times with higher inputs, the roots had two to five times higher concentration of phosphorous. With an increase in water level, the grass took up more nutrients from soil and water. There was a direct and significant correlation between nutrient supply and hydrological regime and plant biomass. The effects of both nutrients and water depth on the nutrient uptake vary with age and need to be studied over a longer period (Sengupta et al., 2004).

From the same experiment, it was observed that the height of *T. maxima* in varying water depths increased from 22.5cm to 101.5cm during the three month-study period. The number of offshoot also increased from 1 to 13 at 20 cm water depth. The biomass of the plants also increased from 0.35 g to 3.2 g in leaves, 0.64 g to 6.2 g in shoots and o.28 g to 5.6 g in roots on dry weight basis (Sengupta et al., 2004). The response of *T. maxima* in the phosphorous (P) accumulation ranged from 0.37 g, 0.74 g and 0.66 g (wet soils) to 1.18g, 5.4g and 3.5 g (at 20 cm depth) in leaves, stems and roots respectively. The N concentration in leaves rose from 1.4% to 2.6%. Nitrogen accumulation ranged from 12.8 mg (wet soil) to 63.4 mg (at 20 cm water depth) in leaves (Sengupta et al., 2004).

Climate Change Adaptation

It has been reported that *T. maxima* can grow in wide range of agro-climatic conditions (Bisht and Ahlawat, 1998). As such, it showed higher efficiency in nutrient uptake in spite of its low nutrient demand per unit dry matter production and diverted significantly greater proportions of dry matter as well as nutrients to below ground tissues (Saxena and Ramakrishnan, 1983). This could be the reason why it is drought resistant. It has the comparative advantage of tolerance to harsh environmental conditions such as steep rocky mountain slopes, shallow soil, drought and high rainfall conditions. It is suitable to grow on wastelands, jhum fallow, as well as in homesteads (Bisht and Ahlawat, 1998). It is frost sensitive and tolerant to neglect (Ramm Botanicals, 2009).

In a study conducted by Khadka (2011) in the Lwang Ghalel Village in India about climate change impact, 65% of the respondents reported that *T. maxima* was the best adapted species for climate change due its fibrous roots. In another case study from Meghalaya in India, there has been significant rise (83%) in tiger grass cultivation during the past three decades and a large portion of land is used for this purpose because these grasses fetched the villagers better price and are least affected by climate change (Lyngdoh and Baishya, 2010).

Other Uses

T. maxima is also found to be of manifold uses other than those presented. The plant itself is used as a weed suppressor (Khisa et al., 1999) and support stake for trailing crops (Rai and Sharma, 1994 in Bhuchar, 2008). It is also ideal for landscape and ornamental purposes (Ramm Botanicals, 2009). Its leaves are used as mulching, roofing materials (Bisht and Ahlawat, 1998) and wrapper for steamed foods (Asia). Its woody culms are used for fuel (Bisht and Ahlawat, 1998; Shankar et al., 2001), reed-pens (Rai and Sharma, 1994 in Bhuchar, 2008) and a support for the cotton wick to offer daily butter lighting in monastery (Namsa et al., 2011). The stems are also used as wall building material (Bisht and Ahlawat, 1998). In the Philippines, its panicles are dyed as carnival costumes and decorative extenders. These other uses of *T. maxima* particularly of that as a fuel can help reduce the dependence of the people on forests for firewoods. It may also qualify as feedstocks for biomass and renewable energy technologies.

T. maxima has also been used as substrate for the cultivation of oyster mushroom, *Pleurotus sp.* Non-composted tiger grass gave higher yield of Bhutan type and gave higher nutrition values as compared to other grasses used like flute reed *(Phragmites karka)* and wild cane *(Saccharum spontaneum Linn.)* (Srijumpa, 2002).

An attractancy test was conducted to explore its role in biological control. One-gram extract of tiger grass per trap exhibited moderate attractive potency to the Oriental fruit fly *(Dacus dorsalis)* which entered 1, 6, 12 and 24 hours into traps in an olfactometer (Areekul et al., 1988). Oriental fruitfly is the second most destructive fruit fly pests in east Asia and the Pacific (Mau and Matin, 2007).

The culms of *T. maxima* also serve as host to a new species of fungi, *Ommatomyces terrestris* (Wang et al., 2000) and a bacteria called *Xanthomonas axonopodis pv. vasculorum.*

Challenges

Extraction of non-timber forest products (NTFPs) like *T. maxima* is an effective conservation strategy to safeguard biological diversity while enhancing rural income. However, excessive harvests may lead to the extinction of species populations or alternatively domestication by the rural people (Shankar, et al., 2001). Overconsumption due to excessive use, because of the high demand in both rural and urban area is making these grasses vulnerable to local extinction (Pandit et al., 2008). Collecting them from their natural habitat may lead to decreased species diversity (Terry and Cunningham, 1993; Karki, 1995, 2001; Wiersum, 1997a; Van Dijik and Wiersum, 1999 all in Sarmah and Arunachalam, 2011). Thus cultivation of *Thysanolaena maxima* should be done in man-made forests and in other land use systems (Sarmah and Arunachalam, 2011) however its susceptibility to fire must be considered (NCVST, 2009). In other places like the Pinatubo area in the Philippines, recruitment of *T. maxima* in the moutains have been hindered due to the chronic human travel and access brought about by tourism (Marier, 2011). Fetalvero et al. (2011) also reported that the culms of *T. maxima* are susceptible to rodent attacks particularly if the undersides are not cleared at least thrice a year.

In India, the gene-pool of tiger grass is depleting due to biotic pressure and developmental activities. Seedling mortality has been recorded due to browsing and grazing, depletion of seed bank, conversion of forest lands for cultivation and construction purposes, large-scale collection of panicles before senescence and dispersal of seeds from brooms. Germplasms have been established to address this concern (Bisht and Ahlawat, 1998).

However, in Queensland, the spread of *T. maxima* from cultivation is a potential threat to riparian vegetation and other damp or shady sites in tropical and sub-tropical regions. Although it is yet to appear in dense stands or cause serious problems, its large tufted growth habit and quick growth rate suggests it has the capacity to outcompete native species in the ground layer (Weed Watch Australia, 2011).

While most studies about *T. maxima* made mention of its resistance to pest, Brunings et al. (2009) observed a leaf spot on tiger grass in 2006. The causal agent of the leaf spot was isolated and characterized morphologically and molecularly as *Exserohilum rostratum* (Dreschsier) Leonard & Suggs. This newly discovered disease could potentially have a dramatic effect on the aesthetic quality and salability of *T. maxima* as a landscape ornamental.

4 TIGER GRASS STUDIES

Rising from the Grassroots. Servañez and Servañez prepared a technical report entitled "The Tiger Grass Industry in Romblon-Rising from the Grassroots". The paper cited data from the Bureau of Agricultural Statistics that the area planted to tiger grass is 238 hectares in the towns of San Agustin and Calatrava. Broom making from tiger grass was introduced by a Japanese volunteer 30 years ago but the progress of the industry was relatively slow. This may be due to the seasonal nature of the resource and the lack of facilities to support the industry. The lack of storage and drying facilities compounds the problem.

In order to address the above issues, several agencies began recognizing this multi-million peso industry. An organization of growers and processors was organized into a cooperative by the Cooperative Development Authority (CDA). The cooperative was granted a loan by the Department of Trade and Industry (DTI) and the Land Bank of the Philippines (LBP) to expand its plantation. The Department of Science and Technology (DOST) had conducted a seminar stressing the need for quality control, uniformity of designs and standards in manufacturing.

Another recent development in the industry was the identification of tiger grass as the community product in the One-Town-One-Product (OTOP) concept espoused by the CASAGANAAN cluster of municipalities which include Calatrava, San Agustin, San Andres, and Santa Maria. In 2003, the local government unit of San Agustin under the mayorship of Hon. Lourdes C. Madrona received a notice from the Japan International Cooperation Agency entitling the municipality to a $50,000 grant for the OTOP concept. The project was implemented in 2004 where a tiger grass processing facility was established. The facility is composed of working and storage building complete with office equipment, 5,000 board foot capacity furnace type kiln dryer and woodworking equipment for the manufacture of broom handles.

A 1,000 square meter lot for the cooperative was donated by Mrs. Norma de la Cruz. The LGU forged a memorandum of agreement with the DOST IV and Forest Products Research and Development Institute (FPRDI) which stated that the latter will provide the blueprints for the dryer and the technical expertise to manage its construction and oversee its commissioning. The facility began operating in 2005.

With this in full functioning, it is expected that the upcoming harvest seasons would bring P5 million to P7 million to the industry. This would come from around the 350,000 brooms that would be produced year round. On top of these, it is expected that an increase in broom prices by at least 10 percent would come as a result of improved design, better handles and quality-assured raw materials. Added benefits would also be realized by the cooperative in the manufacture of broom handles. This may be priced at P700,000 if a broom handle may be made at P2. Added income may also be realized from drying services. At P0.50 per bundle, the cooperative may be able to realize another P500,000 in the future harvest season.

With this, the project stakeholders and people engaged in the industry are hopeful that the tiger grass industry would rise at last from its grassroots and make another name for Romblon as the broom central of the Philippines.

Profile of Tiger Grass Industry in Romblon

The Provincial Science and Technology Center (PSTC) released quick facts based from the information given by DTI about Tiger Grass Broom Industry in the province of Romblon. However, these data were not dated such that there were conflicting figures when cross validated with other literatures. The municipalities of San Agustin and Calatrava were identified as the main production areas with 78 hectares and 20 hectares of land devoted to tiger grass growing, respectively. The estimated production volume of raw tiger grass materials in San Agustin was 15,600 bundles while in Calatrava was 4,000 bundles only. The number of walis tambo produced in San Agustin was 31,500 pieces while in Calatrava was 8,000 pieces.

There were 217 people involved in tiger grass industry, 194 of these were farmers while 23 were processors. In San Agustin, there were 176 people involved in the business where 156 were farmers while 20 were processors. In Calatrava, 41 people were involved in the business, 38 of them were farmers while 3 were processors. The average daily income of a worker in the tiger grass broom industry ranged from P75 to P90 per day. The prevailing price of soft brooms vary across the manner they were bought. Pick-up price ranged from P12 to P15 per piece while retail price ranged from P20 to P30 per piece.

The soft brooms are marketed to NCR and CALABARZON areas (60%), Panay (20%), Mindoro (5%), locally (5%) and to the rest country (10%). In order for the development of this industry to be sustained, the following requirements were worked for: establishment of a common service drying facility, propagation of new seedlings to replenish the aging plantation, introduction of new product designs that are price-competitive and continued skills training to sustain the availability of the labor force.

PSTC estimated that annual production of luway brooms ranges from 200,000 to 250,000 pieces for San Agustin alone. This however, is inconsistent with the quick facts data released by DTI office office (31,500 pieces only). Considering 10 percent contribution in the level of production from Calatrava, San Andres, and Sibuyan would make a total production of 280,000 pieces annually worth P5 million pesos or more depending on the actual factory price. According to PSTC, the demand requirement is almost satisfied, but the people involved in the production aspect of the industry do not enjoy favorable economic environment because of the following reasons: present market channels do not offer favorable returns to the processors because of the presence of too many middlemen dealing on the product, each dealer imposing different level of pricing and make; government assistance to improve the technical aspect of the industry is not readily made available; absence of liberal credit facilities; poor quality of the products; and absence of local policies about the industry.

Tiger Grass Industry in Doña Juana, San Agustin

A comprehensive profiling study done in 2004 about tiger grass industry in Doña Juana, a tiger grass growing barangay in San Agustin, revealed that the barrio devotes about 291.10 hectares to tiger grass production or about 43 percent of the barangay's 681.82 hectares.

Tiger grass in the barangay is commonly planted as undercrop for coconut. Planting density varies from 1 x 1 to 3 x 3 meters depending on land form but most common is the 1.5 x 1.5 meter interval. A hectare of land can be planted with 1,000 to 10,000 hills. A planting density of about 4,500 plants per hectare is commonly observed. Using suckers, tiger grass is planted during the rainy months of July to December. The crop bears flowers after a year of planting.

The industry involves around 236 farmers wherein 214 till their own land while 22 are tenants. Engaged in the industry from 2 to 31 years, these farmers possess an average of 12 years farming experience. The broom processing sector comprises about 105 processors who are mostly farmers at the same time.

Dried tiger grasses are packed for sale by bundling about 100 stalks to form a small bundle. A bigger bundle measuring about a meter in diameter is then completed by tying-up together 30 to 40 small bundles. A big bundle consists of 3,000 to 4,000 dried stalks and weighs about 30 to 36 kilograms.

Two types of brooms are produced based on quality, namely ordinary and special. An ordinary broom binds 20 to 35 dried stalks while special broom wraps 30 to 50 stalks. Broom making peaks on the first quarter getting about 70 percent share of total yearly output. The second quarter contributes 20 percent. The remaining 10 percent is produced in the July to September period.

In 2004, Doña Juana harvested about 7,853 bundles of tiger grass equivalent to nearly 267 metric tons and produced about 1,488,443 brooms. It appears that local processors source out raw materials from other San Agustin barangays or municipalities to sustain the broom-making industry. Market destination includes the local market, Mindoro, Panay, Batangas, Manila and other Luzon areas.

Tiger Grass Industry in San Andres, Romblon

In San Andres, Romblon, tiger grass is usually being propagated and processed into brooms specifically, utility brooms (walis tambo). It all started from a wild variety and improved when cultured. Phil.-Japan Cooperation introduced skills in broom making using this fiber as raw materials. In 2005, 94 family heads produced the raw materials, 40% of them, doubles to broom making which is long in population and the capital involved is short.

As to the performance of this enterprise, sales from raw materials reached P480,000 to P1.5 million pesos from finished brooms marketed to Metro Manila and the nearby Panay provinces. The annual production volume was 1,200 bundles (60,000 finished brooms) where each bundle is 1 meter in circumference. This enterprise is labor intensive, providing more jobs to old and young people.

Two producers of tiger grass in San Andres were identified by the Department of Trade and Industry (DTI). One was the Mari-Norte Development Association which produces straight weave with linear length of 52 centimeters and ½ kilogram net weight per broom wrapped in strips red, blue colored tying plastic. The association was also making brooms for household use. Prices ranged from P20 to P40 pesos per piece, but orders could be made for 60,000 to 80,000 volumes. The SME has a product capacity of 100,000 pieces per year. Another SME was the Romblon Malipayon Development Multi-Purpose Cooperative in Marigondon Sur claiming a broom production capacity of 90,000 pieces per year.

In a project proposal about Techno-Demo Forum on Tiger Grass Technologies, the Department of Science and Technology (DOST-PSTC-Romblon) claimed that tiger grass industry in San Andres, Romblon is said to be already becoming a flourishing industry with 126 hectares of land being devoted for farming. The document reported that there are about 120 farmers engaged in planting and harvesting this crop and manufacturing brooms out of it. Their products are being absorbed by the San Agustin market. The annual tiger grass harvest is estimated to be P1 million pesos. DOST through PCIERD was proposing the purchase of an inflorescence remover developed by researchers at Don Mariano Marcos State University. This technology, if approved is expected to improve the post-harvest processing of the tiger grass as well as its quality. According to the document, tiger grass inflorescence in San Andres is traditionally removed by striking the tiger grass bundles against a tree or some hard objects. This poses risk on the health of the processor who may inhale the scattered pollens. This process also weakens the tiger grass stalks.

Tiger Grass Industry in Marigondon Norte, San Andres, Romblon

One of the most recent and comprehensive studies conducted about the tiger grass industry was the DTI-RSU collaborative study conducted by Fetalvero and Faminial (2010) which profiled the industry in Marigondon Norte in the municipality of San Andres, Romblon.

Marigondon Norte is a tiger grass growing village in San Andres, Romblon. It is located in the northeastern part of the municipality along the lush and rolling mountains of central Tablas. It is bounded by San Agustin in the east and Calatrava in the north. It has an approximate total land area of 2,800 hectares and is home to about 221 households of 1,175 (NSO, 2005) people. It is a typical remote agricultural Philippine village, barely reached by technological breakthroughs because of its rugged trails and far-flung distance (about 14-18km) from the town proper. Although passable to robust motorcycles and sturdy vehicles, transportation remains a major problem because large portions of the service road are still in bad shape and condition.

Tiger grass industry is already a long time economic activity in the locality but its potential has not been fully exploited. Records revealed that a number of interventions were made for the industry to pick up but they were short-lived. It was not until the Department of Trade and Industry (DTI) identified tiger grass as a crop for One Town, One Product (OTOP) program of San Andres that efforts were rekindled to help develop the potential of this industry. Although some data about tiger grass industry in the community were available, these were insufficient to start a strategic and effective intervention. To establish comprehensive baseline data from which developmental efforts are to be anchored, a survey among 100 tiger grass farmers was conducted in the community during the months of December 2009 and January 2010. The list of farmers was provided by the Barangay Council of Mari-Norte but others were identified by key informants including previous barangay officials and Sangguniang Kabataan Chairman. A questionnaire, worded in Filipino, was prepared, face validated by experts from DTI and RSU and pretested among former tiger grass farmers. Vernacular was used in the actual interview.

Farmers' Demographic and Socio-Economic Profiles. Results showed that out of the 100 tiger grass farmers surveyed, 71 were males and 29 were females. Most were married with an average of 4 dependents. The age of farmers ranged from 22 to 79 years old but most were 42 years old. About 80 percent of them were not able to receive a college education and majority spent few years only in high school. Their approximate annual income varied between P2,500 to P130,000 with an average of P20,500 per farmer. The combined annual income of these farmers was P2,683,000 which was usually sourced from tiger grass farming, copra production, poultry, tiger grass processing and nito handicrafts. Most of these farmers were not affiliated with any social organization.

Industry Profiles. There were 86 farmers whose economic activity was concentrated on tiger grass farming alone while 14 doubled to farming and processing. There was some hesitance on the part of most farmers to venture into soft broom making because the process entailed additional labor and they preferred quick cash. The length of their farming experience ranged from 1 to 38 years with an average of 10 ½ years. Most of the broom processors were engaged in the industry for about 20 years already, others for 38 years. Most of the tiger grass farms were owned by farmers but some were tilled by tenants. It was very common to see farms that were 2 km away from homes. Although some farms were situated just beside homes, others were located as far as 5 km away. The farm area estimates ranged from 0.3 to 9 hectares with an average of 1 hectare per farmer. The total farm size was 130.6 hectares distributed in the following locations: Ambunan (39 has.), Hagnaya (37.8 has.), Naruntan (24.75 has.), Lindero (14.3 has.) Hagimit Big (7.75 has.) and Hagimit Small (7 has.).

Farming Practices. Tiger grass farms were also planted with coconut, other rootcrops and palay while others were solely dedicated to tiger grass plantation. The crop calendar usually began with site clearing using the slash and burn (kaingin) technique as early as January or February. By May, the land was ready for the sowing of palay seeds. Around June or July, when the palays were already about a foot tall, tiger grasses were planted alongside with the palays. By September or October, palays were harvested but the tiger grasses were left growing. By January to February, the tiger grasses began producing flowers and by March or April, these were already harvested.

Managing a tiger grass plantation required an estimated annual expenses of P100 to P15,000 depending on farm size. However, the average expense per farmer was estimated at P2,000 a year. The reported cumulative annual expense including farm inputs and farm help was P284,600. Farm help was usually provided for free by family members, neighbors, nephews, cousins and friends. But some paid about P120 to P150 a day per worker.

The ideal months for planting tiger grass were from June to July. There was no day or time preference for planting because the crop can be planted any time and any day within those months. No superstition related to tiger grass planting was reported. The different farm implements used in planting were tara-tara, a sharp rectangular iron attached to a long wooden handle; tagad, a long piece of wood tapered and sharpened in one end; bolo; pala; and piko.

The most common farming practice was to plant 1 to 5 tiger grass hills (seedlings) per hole with intervals 1 m x 1 m intervals. Others were spacing the crops by 1.5 m x 1.5 m and 3 m x 3 m. With this practice, a hectare of land can be planted with 1,000 to 10,000 hills. Weeding and clearing the underside of the plants were factors affecting flowering performance. These were commonly done once a year by most farmers while others were doing this twice or thrice a year. The following harvest and post-harvest practices were observed: cutting the stalks while the panicles were still green and not yet fully mature, sun drying of panicles for three days; and patting sun dried panicles against rocks to shake off the flowers and pollens.

Farm Outputs. The production volume was measured in terms of bundle, a pack of about 100 stalks of cleaned and sun-dried tiger grass panicles. As of last harvest season, the reported average production volume was 600 bundles per farmer or approximately 80,630 bundles for all farmers. In seasons of low production, volume ranged from 3 to 1,500 bundles with an average of 200 bundles per farmer. In seasons of normal produce, volume ranged from 5 to 3,000 bundles with an average of 400 bundles per farmer. And in seasons of high produce, volume ranged from 30 to 5,000 bundles with an average of 500 bundles per farmer.

Two tiger grass products were produced in the locality: dried luway, the material used in making soft brooms; and the soft broom itself (walis tambo). In harvest months, the tendered price for luway ranged from P10 to P35 per bundle. The average price per bundle was P12. During off-peak months, the tendered price increased between P12 to P50 and the average price of each bundle also increased to P20. Price of walis tambo also varied between peak and off peak months ranging from P10 to P60.

Marketing Practices. Farmers commonly sold their produce to luway wholesalers and to luway sales agents. In 2009, the estimated volume of 80,630 bundles were sold to these local agents: Mr. Manasan of Doña Juana (22,200 bundles), Mr. Robert Gabon (20,830 bundles), other agents in Mari-Norte (17,450 bundles), Mr. Gaciles of San Andres (8,650 bundles), agent for Mindoro (4,000 bundles), agent in Mari-Sur (3,600 bundles), agent for Aklan (3,200 bundles) and agent for Odiongan (700 bundles).

The estimated annual income earned by a farmer for dried luway production alone was about P9,500. It was found out that an estimated P1,122,500 income could be realized from this industry representing about 50 percent of the farmers' total annual income estimate which was P2,263,000.

Problems Encountered. Common problems encountered by farmers were lack of financial resource for clearing, labor pay and seedling acquisition; attack of rodents like rats specially when the farms were not cleaned; lack of support from the local government in terms of finding a market; low tendered price for products; poor product quality particularly when it rained during harvest and drying process; and absence of tiger grass processing facilities.

Given proper attention, focus and sustained support, tiger grass production and soft broom processing promise a potential multi-million industry for Barangay Marigondon Norte and for the municipality of San Andres in general. If the industry's 2009 production volume of 80,630 bundles can be maintained or improved, it can generate an estimated revenue of P1M to P2M depending on the prevailing market prices. However, if these raw materials were to be processed into soft brooms, an estimated 241,890 brooms can be produced creating an annual revenue ranging from P3.6M to P7.3M. Figures may be higher if their primitive farming and traditional marketing practices could be improved.

Thus, it was highly recommended that tiger grass farmers in this locality should be reorganized and a comprehensive plan for the tiger grass industry including soft broom processing be prepared. Mature tiger grass technologies were also recommended to be introduced, validated and transferred in the area.

Assumptions of the Study

The following were the assumptions of this investigation:

1. There is no comprehensive data about the tiger grass production potential of the whole province, most are based only on rough estimates;
2. Tiger grass farmers shared common farming practices; and
3. The industry is a promising economic enterprise but hampered by lack of promotional and financial support.

Definition of Terms

Bundle is made up of 100 panicles of cleaned and sun-dried tiger grass flowers.

Inputs refers to the resources utilized in order to propagate, produce or process tiger grass products. This includes the estimated annual farming/processing expenses, farm help and daily wage of farm helpers.

Outputs refers to the farm/broom production in terms of production volume, price of products and estimated annual income from the farming/processing industry.

Marketing practices refers to the activities practiced by the farmers in selling their products.

Panicle refers to a cluster of flowers on a plant attached to an individual stalk.

Tiger grass (luway) is a plant species belonging to grass family that grows to a height of 2 to 3 meters whose long and loose diversely branching flower clusters are processed to produce a utility soft broom locally called walis tambo.

Tiger grass industry refers to any economic activity that involves either the propagation and production of tiger grass or the processing of these grasses into *walis tambo* or both.

5 RESEARCH METHODOLOGY

The research method used is descriptive since the very purpose of the study is to describe the current condition of the tiger grass farmers and/or processors, the tiger grass industry and the problems they encountered. Below is the schematic diagram of the design followed in conducting this study.

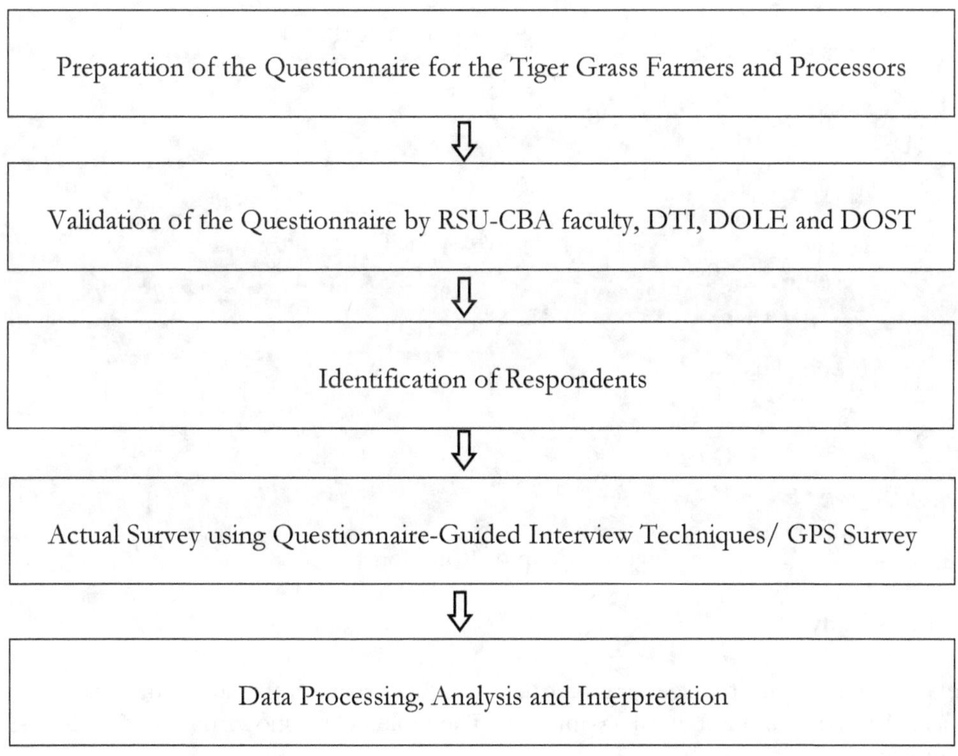

Figure 1. The Research Process

Research Locale and Time of Study

Research Locale. Romblon is one of the archipelagic island provinces located in the MIMAROPA Region which lies north of Boracay Island, east of Mindoro and South of Marinduque. It is bounded on the west by Sibuyan Sea. Tablas is Romblon's largest island composed of nine out of the 17 towns in the province. Located in the central region of Northern Tablas are vast regions of tiger grass villages. These villages adjoin the municipalities of San Andres on the west, Calatrava on the north and San Agustin on the northeast.

San Agustin is previously known as Guintiguian and Badajos. It is a fourth class municipality with a total land area of 8,405 hectares. A mountain range shields San Agustin from southeasterly winds but is completely and barely exposed to northeasterly typhoon which often visits the province. Doña Juana, Binongaan and Hinugusan are located in this town. San Andres is previously known as Salado, Despujol and Parpaguha. It has a total land area of 75.50 square kilometers. It has mountainous and stony areas. Victoria and Mari-Sur are located in this town. Calatrava is previously known as Andagao. It has a total land area of 8,670 hectares constituting 6.39% of Romblon's land area. It is in this town where the village Pagsangahan is located.

Time of Study. The survey was conducted from January 2012 to September 2012. Data processing, GIS mapping and report writing were done from October 2012 to March 2013.

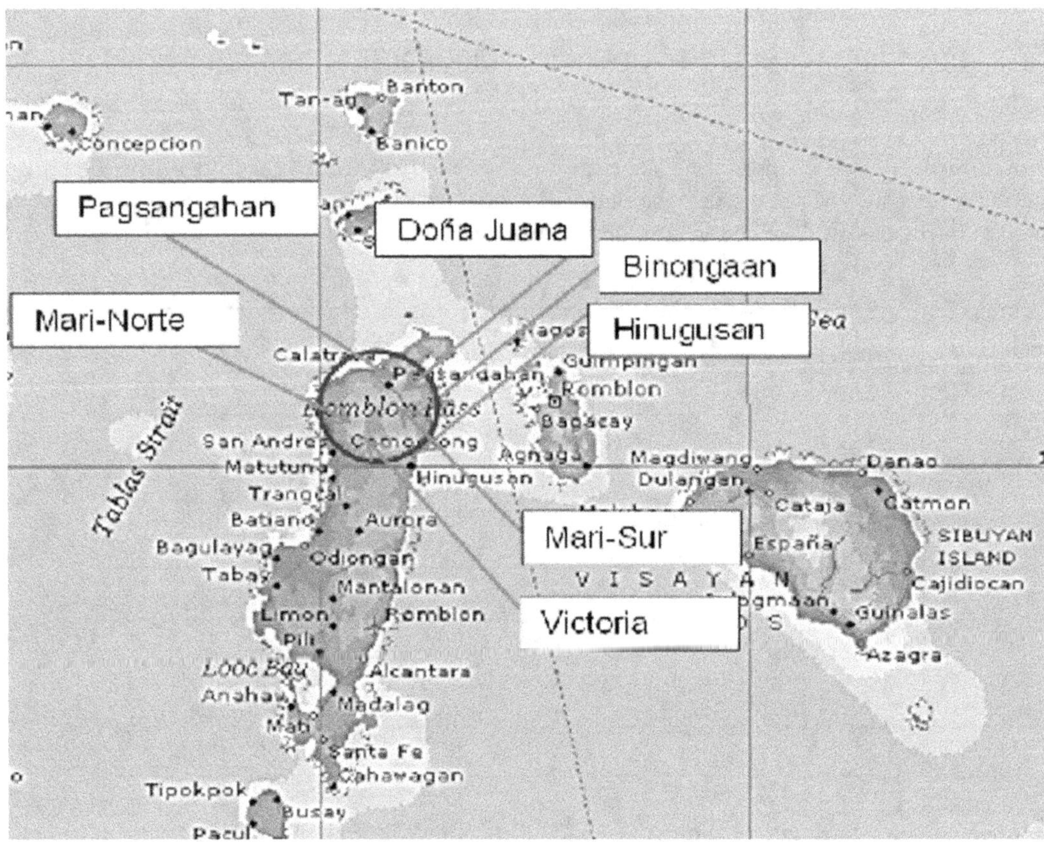

Figure 2. Map of Romblon

The Respondents of the Study

The respondents of the study are the tiger grass farmers and processors in the adjoining villages of Doña Juana, Binongaan, Hinugusan, Victoria, Mari-Sur and Pagsangahan. The table below shows distribution of respondents per village against the estimate.

Table 5. Distribution of respondents per village.

VILLAGE	Estimate	Actual
Victoria	10	7
Mari-Sur	50	29
Pagsangahan	50	30
Doña Juana	90	152
Binongaan	50	56
HInugusan	50	18
TOTAL	300	292

Research Instrument

The data-gathering instrument used in this study was a structured questionnaire which was also used as an interview schedule. It covered pertinent data about the tiger grass farmer/processor profile, tiger grass industry profile and problems encountered relevant to tiger grass industry. The interview schedule was a combination of close and open-ended questions. It was worded in Filipino to ensure that the questions were understood by the respondents. The interview questionnaire was adapted from the developed instrument of Fetalvero and Faminial (2010).

Global Positioning System. This was used in determining the exact coordinates of the tiger grass farms in six villages.

Data Collection Method, Processing and Analysis

The method of data collection was a survey particularly the face-to-face structured interview technique. Enumerators were employed. Possible sources of other information included the following: the barangay captain and secretary and buyers of the tiger grass products. Map coordinates (GPS readings) were determined using the Garmin Foretrex 101.

Results of the interview were systematically encoded with the aid of coding manual and coding sheets. These were processed using the program Microsoft Excel. Data files were converted into an SPSS format (Statistical Packages for Social Sciences). The following statistical measures were used in analyzing the data: frequency count, relative frequency, range, median (when there were extreme values), sum and mean. A GIS software was used in mapping the obtained farm coordinates.

6 PROFILES OF TIGER GRASS FARMERS AND PROCESSORS

Nature of Involvement. As shown in Table 6, there were 163 tiger farmers, 118 farmers and processors and only 11 processors. Most of those doubling as farmers and processors were from Binongaan and Doña Juana (109) while the rest of the respondents from the other villages were only involved in farming activities. Likewise, the 11 processors were also from the two aforementioned villages. It can be implied that these two villages had made tiger grass farming and processing industries as an established economic activity relative to the other villages. This calls for possible intervention in encouraging farmers to process the raw materials into brooms which fetch much higher monetary returns.

Sex. Of the 292 respondents, 58.2% were males while 41.8% were females. These data supported previous reports about the active involvement of the women's sector in tiger grass industries particularly in Binongaan (50%) and Mari-Sur (51.7%) making it a potential livelihood for women in all the villages surveyed. Tiger grass farming activities which includes seedling acquisition, planting, weeding, harvesting and drying are indeed tasks that can be done equally by a woman. Much more, the processing of these dried materials into brooms could become a potential cottage industry for women. The case study of Hanny Gadon (Fetalvero & Faminial, 2010) could attest for the growing numbers of women involved in this kind of industry.

Age. The modal class for the age of the respondents was between 45-54 years old (29.1%). Around 26% of the respondents was above this class while 44.9% was below it. The mean age for the distribution was 46.04 ranging from 17 to 78 years old. The age gap was wide, however the age structure was stable at the middle with an increasing trend from the younger to the middle age and a decreasing trend from the middle to senior age. The trend was evident in all villages surveyed.

Civil Status and Number of Dependents. Most of the respondents were married (82.2%). Interestingly, 8.6% were widow/ers. The industry, although seasonal, has become an alternative source of income for these groups that have dependents to support which averaged to four each respondent and ranged from none to 11.

Socio-Economic Profiles of Tiger Grass Farmers and/or Processors
(see Table 7)

Educational Attainment. As with the previous baseline study among tiger grass farmers (Fetalvero et.al, 2010), lack of adequate education was one of the characteristics of the tiger grass farmers. It can be seen from Table 7 that 24.7% of the respondents did not even finish their elementary education but most did (29.5%). It can be seen too in the table that six respondents (2%) from Doña Juana and Mari-Sur did not receive a formal education at all. However, there were also respondents who reached high school (17.5%), finished high school (16.8%), reached college (4.5%), finished college (2.7%) and finished a vocational course (2.4%). The tiger grass industry therefore is mainly engaged in by those who did not have the opportunity to earn higher levels of education which might be a hindrance in finding a job that pays more. The industry is basically agricultural and practical and does not necessitate a higher level of education, although there is a possibility that informed practices, both in farming and processing technologies, can possibly help improve their yields/production and income.

Table 7. Socio-economic profiles of tiger grass farmers and/or processors.

SOCIO-ECONOMIC PROFILES	VILLAGES						TOTAL	
	Hinugusan	Binongaan	Doña Juana	Victoria	Mari-Sur	Pagsangahan	F	%
EDUCATIONAL ATTAINMENT								
None	0	0	3	0	3	0	6	2.1
Elementary Level	3	10	32	1	9	17	72	24.7
Elementary Graduate	7	19	43	3	10	4	86	29.5
High School Level	2	10	30	1	4	4	51	17.5
High School Graduate	4	7	30	1	3	4	49	16.8
VOCHTECH (Graduated)	1	2	3	1	0	0	7	2.4
College Level	1	4	7	0	0	1	13	4.5
College Graduate	0	4	4	0	0	0	8	2.7
Total	18	56	152	7	29	30	292	100
ESTIMATED ANNUAL INCOME								
Less than P20,000	10	17	69	1	8	17	122	41.8
P20,001 – P39,999	4	16	38	1	15	6	80	27.4
P40,000 – P59,999	2	17	14	0	2	3	38	13.0
P60,000 – P79,999	2	1	16	2	2	1	24	8.2
P80,000 – P99,999	0	1	1	2	0	0	4	1.4
P100,000 and above	0	4	14	1	2	3	24	8.2
Total	18	56	152	7	29	30	292	100
	Range: P1,000 – P500,000			Median: P24,000				
OTHER INCOME SOURCE								
Copra Production	9	14	39	9	28	28	127	43.5
Palay Farming	6	20	21	0	8	1	56	19.2
Fruits and Vegetable production	2	3	6	8	23	4	46	15.8
Rootcrop Production	1	1	6	5	20	9	42	14.4
Manual Labor and Carpentry	0	1	18	1	1	3	24	8.2
Fishing	5	7	8	0	0	0	20	6.8
Retail Store	0	7	11	0	0	1	19	6.5
Salary/Honorarium/Pension	0	3	8	3	0	0	14	4.8
Piggery/Livestock	0	2	6	0	2	2	12	4.1
Corn Farming	0	1	1	4	4	1	11	3.8
Driving	1	3	6	0	0	1	11	3.8
Coconut Broom Making	0	1	6	0	0	0	7	2.4
Tubaan	1	1	4	0	0	0	6	2.1
Mat Weaving/ Nito Handicrafts	0	0	2	0	2	4	4	1.4
MEMBERSHIP IN ORGANIZATIONS								
Yes	8	17	37	5	0	15	82	28.1
No	10	39	115	2	29	15	210	71.9
Total	18	56	152	7	29	30	292	100

Estimated Annual Income. In 2007, the poverty line was P6,195.00/month or P74,340 per year. A five-member family should earn a combined income equal or above that line per month or year in order not to be considered poor (NSCB, 2007). The table shows that on the average, the respondents were earning P24,000 per year (data were based from their income estimate of 2011) which ranged from P1,000 to P500,000. Using the 2007 poverty threshold as basis, their income data revealed that 82.2% of respondents were earning below the poverty line class, 9.6% were earning above it and 8.2% earned within the poverty threshold class.

Other Income Sources. Aside from tiger grass farming and/or processing, the most identified sources of income were copra production (43.5%) and palay farming (19.2%). The table also shows some other sources of the respondents' income like fruits and vegetables production, rootcrops production, manual labor and carpentry, fishing, retail store, salary/honorarium/pension, piggery and livestock, corn farming, driving, coconut broom making, tubaan and mat weaving and nito handicrafts.

Membership in Organizations. A proxy indicator of social empowerment is the respondent's membership in organizations. Only 28.1% of the respondents were involved in community organizations, the other 71.9% were not. Interestingly, the respondents from Pagsangahan (50%) had higher rate of involvement in organizations as compared to the other villages. Among the organizations mostly mentioned during the interview were 4Ps (32), Cooperative (10), TSKI (9), CAR (4), Senior Citizen's Organization (4), Women's organization (2) and Barangay Council.

Profiles of Tiger Grass Farming Industry (See Table 8)

The data on tiger grass farming industry are among the much needed data in planning for a necessary intervention. Table 8 describes the industry as to the years of experience of farmers engaged in tiger grass farming, farm profiles like the distance of farm from home, estimated farm size, land tenure status and land usage. Based from the previous tables, 163 respondents were identified as farmers and 118 were identified as farmers and processors. This is the reason why the summation of frequencies is 281.

Length of Experience as Tiger Grass Farmers. On the average, the respondents have 8 years of farming experience ranging from 0 to 50 years. Around 29% of the farmers were engaged in the industry for less than 5 years only. There seems to be a growing interest in the industry as shown by the trend (increasing number of farmers) for the past 19 years.

Distance of Farm from Home. The average distance of farm from home was 3 km which ranged from 0 to 10 km. Most of the farms surveyed were indeed far as mountains were to be crossed before getting there.

Estimated Size of Tiger Grass Farm. Based on the respondents' estimates, the total farm size in these three barangays was 410.62 hectares. Largest farm areas were in Doña Juana (215.52-ha), followed by Binongaan (88.50-ha), Hinugusan (33.60-ha), Pagsangahan (35.25-ha), Hinugusan (33.6-ha), Mari-Sur (30.25-ha) and Victoria (7.50-ha). In 2010, Fetalvero & Faminial reported that in barangay Mari-Norte, there was about 130.6 hectares of land planted with luway. These values altogether show that there are about 541.22 hectares of tiger grass plantations in Northern Tablas.

Based on the interviews conducted, the village of Lusong which is situated between Hinugusan and Binongaan, is also a tiger grass growing village and was reported to have more production than Hinugusan. The inner and mountainous sitios of Doña Juana, namely Dagasdas, Santa Ana, Ube and Mamayang Kahoy, were also reported to have vast areas of tiger grass plantations. But survey in the area was aborted due to security reasons.

In Doña Juana, the identified tiger grass sitios were Centro, Tinigbana and Ilaya. In Pagsangahan, the identified sitios planted with tiger grass were Centro, Ambunan, Kawa-Kawa and Malabuyo. In Victoria, the identified sitios were Centro and Pader. In Mari-Sur, the identified sitios were Agbalagon, Balikawang, Gin-uslob, Hacienda and Limon. Fetalvero and Faminial (2010) reported that in Mari-Norte the sitios planted with tiger grass were Ambunan, Hagnaya, Naruntan, Hagimit Big, Lindero and Hagimit Small.

Table 8. Tiger grass farming industry profile.

INDUSTRY PROFILES	VILLAGES						TOTAL	
	Hinugusan	Binongaan	Doña Juana	Victoria	Mari-Sur	Pagsangahan	F	%
LENGTH OF EXPERIENCE AS TIGER GRASS FARMER								
Less than 5	11	23	28	4	6	9	81	28.8
5 – 9	4	13	31	3	5	5	61	21.7
10 – 14	2	4	33	0	9	5	53	18.9
15 – 19	1	5	12	0	1	2	21	7.5
20 – 24	0	2	20	0	1	3	26	9.3
25 – 29	0	2	6	0	2	2	12	4.3
30 and above	0	2	16	0	5	4	27	9.6
Total	18	51	146	7	29	30	281	100
	Range: 0 - 50		Median: 8					
DISTANCE OF FARM FROM HOME (km)								
2 and below	5	18	63	6	25	20	137	48.8
2.1 – 4	10	10	17	1	1	8	47	16.7
4.1 – 6	0	15	22	0	3	1	41	14.6
6.1 – 8	3	2	19	0	0	1	25	8.9
More than 8	0	5	25	0	0	0	30	10.7
No response	0	1	0	0	0	0	1	3.6
Total	18	51	146	7	29	30	281	100
	Range: 0 - 10		Median: 3					
ESTIMATED SIZE OF TIGER GRASS FARM (in hectares)								
Less than 0.05	0	1	17	0	5	1	25	9.0
0.5 – 0.99	3	11	34	3	10	12	73	26.0
1 – 1.99	6	23	54	3	6	11	103	36.7
2 – 2.99	5	8	19	1	7	3	43	15.3
3 – 3.99	2	2	9	0	1	2	16	5.7
4 and above	2	6	12	0	0	1	21	7.5
No response	0	0	1	0	0	0	1	0.36
Total	18	51	146	7	29	30	281	100
Range	0.5 – 5	0.25 – 7	0 – 8	0.5 – 2	0.25 – 3	0.25 – 5.50	0 – 8	
Median	1.75	1	1	1	0.75	1	1	
Sum	33.6	88.50	215.52	7.50	30.25	35.25	410.62	
LAND TENURE STATUS								
Owner	4	24	68	2	9	14	121	43.1
Tenant	1	21	54	3	17	10	106	37.7
Renting	3	5	15	1	0	5	29	10.3
CARP, IP or PWD Beneficiary	8	1	3	0	0	0	12	4.3
Others	1	0	2	0	0	0	3	1.1
No response	1	0	4	1	3	1	10	3.6
Total	18	51	146	7	29	30	281	100
LAND USE								
Planted with tiger grass	3	2	32	0	0	3	40	14.2
Planted with palay and then *luway*	8	21	12	7	26	20	94	33.5
Planted with *luway*, coconut and other undercrops	12	41	108	7	26	20	214	76.2
Others	1	10	6	0	0	1	18	6.4

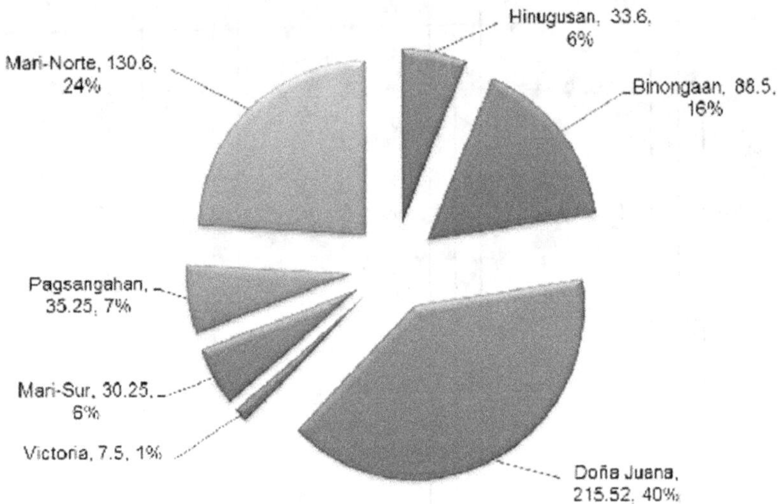

Figure 3. Pie-graph showing the size of tiger grass farm in Northern Tablas.

Land Tenure Status. Most of the farmers were owners (43.1%), if not tenants (37.7%), of the farm. There were also those who were renting (10.3%) and CARP beneficiaries (4.3%).

Land Use. As to the land use, 76.2% of the respondents reported that their farms were planted with luway, coconut and other undercrops; 33.5% planted their farms with palay first and then luway while 14.2% said that their farms were solely intended for tiger grass plantation.

Farming Practices

Farming calendar and practices in these six villages were almost similar with those discussed in Fetalvero and Faminial (2010) which was cited in this investigation (see p. 28). The most common interval in planting was 1 meter x meter (44.8%) as shown in Table 9.

Table 9. Farming practices.

FARMING PRACTICE	FREQUENCY	RELATIVE FREQUENCY
0.5 x 3 span	4	1.4
1 x 1 span	13	4.6
1 x 1 meter	126	44.8
1 x 2 meters	3	1.1
1 x 3 meters	4	1.4
1.5 x 1.5 meters	19	6.8
1.5 x 2 meters	3	1.1
2 x 2 meters	12	4.3
No response	97	34.5
Total	281	100.0

Farm Inputs

Literatures reported that tiger grass farming requires minimal inputs. Shown in Table 10 are the estimated annual farm expense, farm and human resource needs.

Estimated Annual Farm Expense. As seen from Table 10, around 54% of the respondents reported expenditures in establishing and maintaining the farm. But a large percentage (45.9%) did not spend any peso at all. The average annual farm expenditure was P5,000 and ranged from P150 to P100,000 per farmer.

Table 10. Farm inputs.

FARM INPUTS	VILLAGES						TOTAL	
	Hinugusan	Binongaan	Doña Juana	Victoria	Mari-Sur	Pagsangahan	F	%
ESTIMATED ANNUAL FARM EXPENSE								
Less than P1,000	0	3	8	0	3	0	14	5.0
P1,000 – P1,999	2	3	7	0	4	2	18	6.4
P2,000 – P2,999	3	2	9	1	1	3	19	6.8
P3,000 – P3,999	0	5	4	1	2	2	14	5.0
P4,000 – P4,999	1	0	4	0	0	1	6	2.1
P5,000 – P5,999	0	10	21	1	1	2	35	12.5
P6,000 and above	3	15	26	1	1	0	46	16.4
No expense	9	13	67	3	17	20	129	45.9
Total	18	51	146	7	29	30	281	100.
	Range: P150 – P100,000		Median: P5,000					
FARM NEEDS								
Fertilizer	1	1	5	0	0	0	7	2.5
Pesticide	1	0	7	0	0	0	8	2.8
Farm Implements	18	51	135	7	29	30	270	96.1
Others	0	0	2	0	0	0	2	0.7
NEED OF FARM WORKERS								
Yes	6	27	44	4	11	6	98	34.9
No	12	24	102	3	18	24	183	65.1
Total	18	51	146	7	29	30	281	100
RELATIONSHIP TO FARM WORKERS								
Relatives	3	7	23	3	2	4	42	42.9
Neighbors	0	7	5	0	7	0	19	19.4
Family Members	1	10	10	1	2	0	24	24.4
Friends	2	3	6	0	0	2	13	13.3
Total	6	27	44	4	11	6	98	100
ARE THE FARM WORKERS PAID?								
Yes	6	26	42	4	10	6	94	95.9
No	0	1	2	0	1	0	4	4.1
Total	6	27	44	4	11	6	98	100
DAILY WAGE OF FARM WORKERS								
P100	0	0	1	2	4	0	7	7.4
P120	2	1	1	2	1	0	7	7.4
P150	2	24	15	0	4	1	46	48.9
P160	0	0	4	0	0	0	4	4.3
P165	0	0	1	0	0	0	1	1.1
P170	0	0	6	0	0	0	6	6.4
P180	0	0	2	0	0	0	2	2.1
P200	2	1	12	0	1	5	21	22.3
Total	6	26	42	4	10	6	94	100

Farm Needs. Farm implements like tara-tara (a sharp rectangular iron attached to a long wooden handle), bolo and tagad (a sharp and pointed wooden pole) were used by 96.1% of the farmers. Pesticide (2.8%) and fertilizers (2.5%) were among the farm needs reported.

Human Resource Needs. Farm help is not needed anymore by 65.1% of these farmers. This implies that the tiger grass farming industry can be established solely by one person. However, the rest of the respondents sought help from any of the following: relatives (42.9%), family members (24.4%), neighbors (19.4%) and friends (13.3%). Among those who needed farm help, only four did not pay for labor. The rest of the 94 farmers pegged daily wage of farm workers at an average of P150 per day (48.9%).

Farm Outputs

Farm outputs in terms of production volume, pricing, and income generated from tiger grass farming are presented in this section. Production was measured in bundles and compared during low, normal and high volumes. A bundle is made up of 100 panicles of cleaned and sun-dried tiger grass flowers.

Production volume. During low production, the harvest ranged from 1 to 25,000 bundles with an average of 120 bundles (Table 11). During normal production, harvest ranged from 2 to 50,000 bundles with an average of 200 bundles per farmer. During high production, volume ranged from 3 to 80,000 bundles with an average of 300 bundles per farmer.

Tendered price. As shown in Table 12, price of bundles varied across peak months (tiger grass harvest season) which ranged from February to May and off peak months from October to December. The cost of one bundle during peak months ranged from P10 to P80 with an average of P25. But during off-peak months, one bundle costs more averaging to P52 each and ranged from P15 to P150. Price depends on the availability of the supply.

Number of bundles sold. During the last harvest season, the average number of bundles sold ranged from 2 to 7,500 bundles and averaged to 150 bundles (refer to Table 13). For a possible market channel study, the list of buyers of dried luway was provided in Appendix B.

Estimated Annual Income from Tiger Grass Production. Table 14 presents the estimated annual income from tiger grass production activity which ranged from P50 to P187,500. On the average, each farmer earned P3,750, a figure which was much lower than the previously reported income of P9,000 per farmer (Fetalvero & Faminial, 2010). This might be due to a large percentage of the respondents (30.2%) who refused to answer this question because they themselves processed the flowers into softbrooms. Others were hesitant because of fear that they might be reported to the BIR for taxing purposes.

Table 11. Production volume in bundles.

PRODUCTION VOLUME (in bundles)	VILLAGES						TOTAL	
	Hinugusan	Binongaan	Doña Juana	Victoria	Mari-Sur	Pagsangahan	F	%
LOW PRODUCTION								
Less than 500	4	40	121	3	23	27	218	77.6
500 – 999	1	1	13	1	1	1	18	6.4
1,000 – 1,499	1	0	1	0	0	1	3	1.1
1,500 – 1,999	0	0	0	0	1	0	1	0.4
2,000 – 2,499	1	0	1	0	0	0	2	0.7
2,500 and above	1	1	3	0	0	0	5	1.8
No response	10	9	7	3	4	1	34	12.1
Total	18	51	146	7	29	30	281	100
	Range: 1-25,000		Median: 120		Sum: 25,000			
NORMAL PRODUCTION								
Less than 500	4	37	111	3	22	25	202	71.9
500 – 999	0	6	17	1	2	3	29	10.3
1,000 – 1,499	0	1	8	0	0	1	10	3.6
1,500 – 1,999	1	0	3	0	1	0	5	1.8
2,000 – 2,499	1	0	1	0	0	0	2	0.7
2,500 and above	2	1	3	0	0	0	6	2.1
No response	10	6	3	3	4	1	27	9.6
Total	18	51	146	7	29	29	281	100
	Range: 2 – 50,000		Median: 200		Sum: 192,183			
HIGH PRODUCTION								
Less than 500	3	31	97	2	22	23	178	63.3
500 – 999	1	8	24	1	3	4	41	14.6
1,000 – 1,499	0	1	10	2	3	1	17	6.0
1,500 – 1,999	0	0	6	0	0	0	6	2.1
2,000 – 2,499	0	2	5	0	1	1	9	3.2
2,500 and above	4	4	4	0	0	0	12	4.3
No response	10	5	0	2	0	1	18	6.4
Total	18	51	146	7	29	30	281	100
	Range: 3 – 80,000		Median: 300		Sum: 251,022			

Table 12. Tendered price per bundle of luway.

PRICE PER BUNDLE	VILLAGES						TOTAL	
	Hinugusan	Binongaan	Doña Juana	Victoria	Mari-Sur	Pagsangahan	F	%
PEAK MONTHS								
P10 – P19	0	8	21	1	18	16	64	22.8
P20 – P29	1	15	36	1	5	5	63	22.4
P30 – P39	4	8	45	1	2	5	65	23.1
P40 – P49	1	7	13	0	0	3	24	8.5
P50 and above	2	6	24	0	0	0	32	11.4
No response	10	7	7	4	4	1	33	11.7
Total	18	51	146	7	29	30	281	100
	Range: 10 - 80		Median: 25					
OFF-PEAK MONTHS								
P10 – P19	0	0	1	0	0	0	1	0.4
P20 – P29	0	0	5	0	5	3	13	4.6
P30 – P39	1	1	4	0	4	6	16	5.7
P40 – P49	0	1	6	3	6	13	29	10.3
P50 and above	7	43	124	0	11	7	192	68.3
No response	10	6	6	4	3	1	30	10.7
Total	18	51	146	7	29	30	281	100
	Range: 15 - 150		Median: 52					

Table 13. Number of bundles sold during the last harvest season.

NUMBER OF BUNDLES SOLD	VILLAGES						TOTAL	
	Hinugusan	Binongaan	Doña Juana	Victoria	Mari-Sur	Pagsangahan	F	%
100 and below	1	11	45	2	16	14	89	31.7
101- 300	2	11	29	0	7	8	57	20.3
301 – 600	1	3	12	0	2	5	23	8.2
601- 900	0	1	1	0	0	1	3	1.1
Above 900	3	12	6	1	1	1	24	8.5
No response*	11	13	53	4	3	1	85	30.2
Total	18	51	146	7	29	30	281	100
	Range: 2 – 7,500		Median: 150					

* Other farmers do not sell their dried luway flowers because they process these flowers into brooms by themselves. Others refused to answer because they were afraid that they would be reported to the BIR.

7 PROFILES OF TIGER GRASS PROCESSING INDUSTRY

Profiles of Tiger Grass Processing Industry

Length of Engagement in the Entrepreneurial Activity. Of the 292 respondents, 11 are processors and 118 doubled as farmers and processors that is why the summation of frequency is 129. The average length of experience as tiger grass processors was 10 years, ranging from 1 to 34 years (Table 15). It can be further seen from the table that around 23% of the respondents had just ventured into the business in less than 5 years which is indicative of a growing interest in tiger grass processing industry.

Tiger Grass Processing Inputs. As shown in Table 16, the estimated average capital needed in processing tiger grass flowers into soft brooms was P3,000. But this differed from processor to processor as the business capital ranged from P250 to P500,000 per farmer. These are usually spent for the needed materials like plastic rattan, tying wires and stickers. This is also dependent on the volume of dried flowers to be processed. In an interview with the respondents, almost all of the farmers took this capital amount from their savings. But there were few who loaned money from lenders with 10% monthly interest.

pioneered in duster production and it was sold from P20 during peak season and P45 during off-peak season.

Estimated Annual Income from Tiger Grass Processing. It can be seen from Table 19 that a processor can earn an income of P17,500 a year in making soft brooms out of tiger grass flowers. This is relatively higher as compared to the annual income that can be generated by selling the dried flowers alone (P3,750). Although it was expected that processing can increase the value of the tiger grass flowers, the reported data in this study must be taken with caution because of the problem in reporting the income from tiger grass production. As mentioned earlier, around 30% of the respondents were reluctant in reporting the real figures with the fear that the survey was commissioned by the Bureau of Internal Revenues. Some others underestimated their income while others refused to give any information anxious that they would be removed from the list of DSWD's indigents. Should these respondents honestly reported their income, then the data could have been different.

Marketing and Packaging Practices

As shown in Table 20, softbrooms were usually purchased by wholesalers on site (47.3%), others were delivered to customers and market channels (13.2%), some were sold directly from house to house (8.5%) and a few were displayed in stalls (1.5%).

Table 14. Estimated annual income from tiger grass production.

ANNUAL INCOME	VILLAGES						TOTAL	
	Hinugusan	Binongaan	Doña Juana	Victoria	Mari-Sur	Pagsangahan	F	%
P5,000 and below	1	15	61	2	21	21	121	43.1
P5,001 to P10,000	3	7	16	0	2	3	31	11.0
P10,001 to P20,000	0	3	8	0	2	4	17	6.0
P20,001 to P30,000	1	6	3	1	1	1	13	4.6
P30,001 to P40,000	0	0	2	0	0	0	2	0.7
Above P40,000	2	7	3	0	0	0	12	4.3
No response*	11	13	53	4	3	1	85	30.2
Total	18	51	146	7	29	30	281	100
	Range: 50 – 187,500		Median: 3,750					

* Other farmers do not sell their dried luway flowers because they process these flowers into brooms by themselves. Others refused to answer because they were afraid that they would be reported to the BIR.

Table 15. Length of experience as tiger grass processors.

LENGTH OF EXPERIENCE AS TIGER GRASS PROCESSORS	VILLAGES						TOTAL	
	Hinugusan	Binongaan	Doña Juana	Victoria	Mari-Sur	Pagsangahan	F	%
Less than 5	1	10	16	2	2	0	30	23.3
5 – 9	1	3	18	0	0	0	22	17.1
10 – 14	0	4	16	0	0	0	20	15.5
15 – 19	0	3	12	0	0	0	15	11.6
20 – 24	0	2	17	0	0	0	19	14.7
25 – 29	0	2	5	0	0	0	7	5.4
30 and above	0	0	9	0	2	1	12	9.3
No response	0	4	0	0	0	0	4	3.1
Total	2	28	92	2	4	1	129	100
	Range: 1-34		Median: 10					

Table 16. Tiger grass processing inputs.

AMOUNT OF CAPITAL	VILLAGES						TOTAL	
	Hinugusan	Binongaan	Doña Juana	Victoria	Mari-Sur	Pagsangahan	F	%
P1,000 and less	2	4	31	0	2	0	39	30.2
P1,001 to P5,000	0	9	32	1	2	1	45	34.9
P5,001 to P10,000	0	2	16	0	0	0	18	14.0
P10,001 to P15,000	0	0	4	0	0	0	4	3.1
More than P15,000	0	7	9	1	0	0	17	13.2
No response	0	6	0	0	0	0	6	4.7
Total	2	28	92	2	4	1	129	100
	Range: 250 – 500,000		Median: 3,000					

Table 17. Prices of regular brooms during peak and off-peak months.

PRICE OF REGULAR BROOM	VILLAGES						TOTAL	
	Hinugusan	Binongaan	Doña Juana	Victoria	Mari-Sur	Pagsangahan	F	%
PEAK MONTHS								
P20 and below	0	4	13	0	0	0	17	13.2
P21 – P30	0	14	56	0	1	0	71	55.0
P31 – P40	0	2	21	1	0	1	25	19.4
P41 – P50	0	2	0	1	3	0	6	4.7
P51 – P60	1	0	0	0	0	0	1	0.8
P61 – P70	0	0	1	0	0	0	1	0.8
Above P70	0	1	1	0	0	0	2	1.6
No response	1	5	0	0	0	0	6	4.7
Total	2	28	92	2	4	1	129	100
	Range: 12-80		Median: 25					
OFF-PEAK MONTHS								
P20 and below	0	0	5	0	0	0	5	3.9
P21 – P30	0	17	41	0	1	0	59	45.7
P31 – P40	0	4	34	2	0	0	40	31.0
P41 – P50	0	0	8	0	2	0	10	7.8
P51 – P60	0	1	3	0	1	0	5	3.9
P61 – P70	0	0	0	0	0	0	0	0.0
Above P70	1	1	1	0	0	1	4	3.1
No response	1	5	0	0	0	0	6	4.7
Total	2	28	92	2	4	1	129	100
	Range: 12 - 90		Median: 30					

Table 18. Prices of jumbo brooms during peak and off-peak months.

PRICE OF JUMBO BROOM	VILLAGES						TOTAL	
	Hinugusan	Binongaan	Doña Juana	Victoria	Mari-Sur	Pagsangahan	F	%
PEAK MONTHS								
P21 – P30	0	0	1	0	0	0	1	6.3
P31 – P40	0	0	0	0	1	0	1	6.3
P41 – P50	1	0	2	1	0	0	4	25.0
P51 – P60	0	2	3	1	0	0	6	37.5
P61 – P70	0	0	0	0	0	0	0	0
Above P70	0	2	1	0	0	1	4	25.0
Total	1	4	7	2	1	1	16	100
	Range: 30 - 100		Median: 60					
OFF-PEAK MONTHS								
P31 – P40	0	0	1	0	1	0	2	12.5
P41 – P50	1	0	1	1	0	0	3	18.8
P51 – P60	0	0	1	1	0	0	2	12.5
P61 – P70	0	1	1	0	0	0	2	12.5
Above P70	0	3	3	0	0	1	7	43.8
Total	1	4	7	2	1	1	16	100
	Range: 35 - 120		Median: 67.50					

Table 19. Estimated annual income from tiger grass processing.

INCOME	VILLAGES						TOTAL	
	Hinugusan	Binongaan	Doña Juana	Victoria	Mari-Sur	Pagsangahan	F	%
P5,000 and below	2	3	11	0	1	0	17	13.2
P5,001 to P15,000	0	7	24	0	0	0	31	24.0
P15,001 to P25,000	0	5	11	0	1	0	17	13.2
P25,001 to P35,000	0	2	7	1	2	0	12	9.3
P35,001 to P45,000	0	0	2	0	0	0	2	1.6
P45,001 to P55,000	0	1	1	0	0	0	2	1.6
Above P55,000	0	6	10	1	0	1	18	14.0
No response*	0	4	26	0	0	0	30	23.3
Total	2	28	92	2	4	1	129	100
	Range: 900 – 420,000		Median: 17,500					

* Other farmers do not sell their dried luway flowers because they process these flowers into brooms by themselves. Others refused to answer because they were afraid that they would be reported to the BIR.

Table 20. Marketing practices of products.

MARKETING PRACTICES (n=129)	FREQUENCY	RELATIVE FREQUENCY
Buyers purchased products on-site.	61	47.3
Delivered to customers and market channels.	17	13.2
Direct selling (house to house)	11	8.5
Pre-ordered.	2	1.5
Display in stalls.	2	1.5

In Table 21, around 54% of the processors were observing quality control of their works like careful and tight stitching of well-cleaned and well-dried panicles. They were also improving the stockiness or thickness of the brooms (29.5%), others were adopting labels like "Tablas premium" or "Made in Baguio" (9.3%) while some were focusing on the design of the stitches and handles (7.8%).

Table 21. Product packaging.

PACKAGING PRACTICES (n=129)	FREQUENCY	RELATIVE FREQUENCY
Quality control (carefully and tightly stitched; well-dried; well-cleaned)	69	53.5
Stockiness (bristles)	38	29.5
Label (sticker, Baguio made)	12	9.3
Design (handle and stitching)	10	7.8

As shown in Table 22, the large percentage of market for softbrooms were the wholesalers in Doña Juana (53.5%). The bulk of these brooms stored in Doña Juana warehouses are exported out of the province and sold in Metro Manila. Other buyers of the products were also identified to be coming from Manila, Kalibo, Iloilo, Capiz, Binongaan, Batangas, Odiongan, Bataan, Masbate, Looc, San Agustin, Romblon, Mindoro, Antique, Santa Maria, San Andres, Cavite and Buli.

Table 22. Identified market for softbrooms.

MARKET	VILLAGES						TOTAL	
	H	B	DJ	V	MS	P	F	%
Dona Juana	1	9	59				69	53.5
Manila	1	2	5	1		1	10	7.8
Kalibo			9				9	7.0
Iloilo	1		6				7	5.4
Capiz		1	6				7	5.4
Binongaan		6			1		7	5.4
Batangas			6				6	4.7
Odiongan			3	2			5	3.9
Bataan			5				5	3.9
Masbate		1	2				3	2.3
Looc			2	1			3	2.3
San Agustin			1		1		2	1.6
Romblon		1	1				2	1.6
Mindoro		1				1	2	1.6
Antique			2				2	1.6
Santa Maria			1				1	0.8
San Andres			1				1	0.8
Cavite		1					1	0.8

Problems Encountered by Farmers and Processors

As shown in the Table 23, the most common problems reported by the farmers were infestation of pests and animals (37.7%), theft (35.2%), attack of rodents and monkeys (27.8%), climate change (22.1%). Some minor problems reported include weeds (12.1%), lack of financial resource in clearing, labor pay and seedling acquisition (7.8%), tending the farm is tiresome (5.3%), typhoon (4.3%), farm is distant (1.8%) and bush fire (1.1%).

Table 23. Problems encountered by tiger grass farmers.

PROBLEMS ENCOUNTERED	FREQUENCY	RELATIVE FREQUENCY
FARMERS (n=281)		
Infestation of pests and animals	106	37.7
Theft	99	35.2
Attack of rodents and monkeys	78	27.8
Climate change (drought, too much rain affecting plant's growth and harvest)	62	22.1
Weeds	34	12.1
Lack of financial resource in clearing, labor pay, and seedling acquisition	22	7.8
Tending the farm is tiresome	15	5.3
Typhoon	12	4.3
Farm is distant	5	1.8
Bush fire	3	1.1

Table 24. Problems encountered by tiger grass processors.

PROBLEMS ENCOUNTERED	FREQUENCY	RELATIVE FREQUENCY
PROCESSORS (n=129)		
Physically tiresome	62	48.1
Lacks financial resource	33	25.6
Rain (molds attack the stocked flowers)	30	23.3
Low tendered price.	8	6.2
Lack of materials (plastic, and dried flowers)	7	5.4
Difficulty in removing pollens	4	3.1
Difficulty in finding a market.	3	2.3
Products are rejected by buyers	2	1.6

Processors also identified some problems related to soft broom making as shown in Table 24. These were the physically tiresome task of processing (48.1%), lacks financial resource (25.6%), rain which triggers mold attack (23.3%), low tendered price (6.2%), lack of materials for processing (5.4%), difficulty in removing pollens (3.1%), difficulty in finding a market (2.3%) and products are rejected by buyers (1.6%).

8 TIGER GRASS RESOURCE MAPS

The following GPS readings were generated using the Garmin Foretrex 101. Only the subsequent farms were actually surveyed because of the mountainous and rugged terrains. Farms adjacent to each other were taken as one in determining the GPS coordinates. Only one or two points were taken in sloping farms that were difficult to survey.

Hinugusan

Table 25. GPS coordinates of tiger grass farms in Hinugusan.

FARM	POINTS	NORTH LATITUDE	EAST LONGITUDE
Farm 1	1	12°30'38.0"	122°05'27.0"
	2	12°30'38.1"	122°05'26.8"
	3	12°30'33.9"	122°05'24.8"
	4	12°20'35.6"	122°05'24.1"
Farm 2	1	12°30'39.0"	122°05'18.3"
	2	12°30'39.9"	122°05'18.6"
	3	12°30'41.3"	122°05'17.0"
	4	12°30'40.1"	122°05'16.8"
Farm 3	1	12°30'41.4"	122°05'15.8"
	2	12°30'41.6"	122°05'15.1"
	3	12°30'44.4"	122°05'14.9"
	4	12°30'44.8"	122°05'16.2"
Farm 4	1	12°30'48.1"	122°05'09.3"
	2	12°30'46.7"	122°05'08.0"
	3	12°30'46.7"	122°05'07.7"
	4	12°30'47.7"	122°05'09.9"
Farm 5	1	12°30'48.9"	122°05'11.0"
	2	12°30'49.9"	122°05'10.9"
	3	12°30'48.6"	122°05'12.6"
	4	12°30'49.0"	122°05'12.8"
Farm 6	1	12°30'32.2"	122°05'34.0"
	2	12°30'31.4"	122°05'35.0"
	3	12°30'30.6"	122°05'34.2"
	4	12°30'32.8"	122°05'33.3"
	5	12°30'32.9"	122°05'32.7"
Farm 7	1	12°29'25.1"	122°06'40.4"
	2	12°29'24.9"	122°06'39.5"
	3	12°29'27.7"	122°06'39.3"
	4	12°29'28.5"	122°06'39.5"
Farm 8	1	12°30'15.3"	122°06'39.6"
	2	12°30'15.6"	122°06'38.5"

FARM	POINTS	NORTH LATITUDE	EAST LONGITUDE
	3	12°30'16.5"	122°06'39.5"
	4	12°30'18.9"	122°06'36.2"
	5	12°30'17.7"	122°06'38.1"
Farm 9	1	12°30'24.5"	122°06'32.3"
	2	12°30'24.2"	122°06'31.8"
	3	12°30'24.3"	122°06'31.3"
	4	12°30'22.5"	122°06'30.9"
	5	12°30'22.2"	122°06'29.7"
	6	12°30'24.2"	122°06'29.6
Farm 10	1	12°30'28.2"	122°06'23.3"
	2	12°30'27.9"	122°06'23.7"
	3	12°30'28.0"	122°06'23.4"
Farm 11	1	12°30'20.6"	122°06'43.4"
	2	12°30'20.3"	122°06'43.3"
	3	12°30'20.8"	122°06'45.1"
	4	12°30'20.6"	122°06'44.9"
Farm 12	1	12°29'42.8"	122°06'30.7"
	2	12°29'43.1"	122°06'25.6"
	3	12°29'43.9"	122°06'29.6"
	4	12°29'43.5"	122°06'31.1"
Farm 13	1	12°29'48.7"	122°06'36.9"
	2	12°29'48.6"	122°06'36.4"
	3	12°29'49.0"	122°06'35.8"
	4	12°29'49.1"	122°06'36.2"
Farm 14	1	12°29'49.2"	122°06'39.2"
	2	12°29'48.4"	122°06'38.6
	3	12°29'48.3"	122°06'39.9"
Farm 15	1	12°29'27.7"	122°06'15.6"
	2	12°29'26.8"	122°06'14.7"
	3	12°29'26.4"	122°06'14.8"
	4	12°29'26.0"	122°06'15.6"
	5	12°29'27.6"	122°06'14.5"
Farm 16	1	12°29'37.8"	122°05'45.7"
	2	12°29'37.9"	122°05'46.5"
	3	12°29'38.1"	122°05'46.3"
	4	12°29'38.3"	122°05'46.0"
Farm 17	1	12°29'40.3"	122°05'45.6"
	2	12°29'41.5"	122°05'45.7"
	3	12°29'40.6"	122°05'46.1"
Farm 18	1	12°29'37.9"	122°05'49.3"
	2	12°29'37.3"	122°05'49.5"
Farm 19	1	12°29'37.5"	122°05'49.7"
Farm 20	1	12°29'27.3"	122°05'40.8"
	2	12°29'26.9"	122°05'41.5"
	3	12°29'27.1"	122°05'41.7"
	4	12°29'27.3"	122°05'40.6"
Farm 21	1	12°29'21.5"	122°05'58.3"
	2	12°29'20.6"	122°05'59.2"
Farm 22	1	12°29'34.7"	122°0631.8"
	2	12°29'33.5"	122°06'32.1"
	3	12°29'34.0"	122°06'33.9"

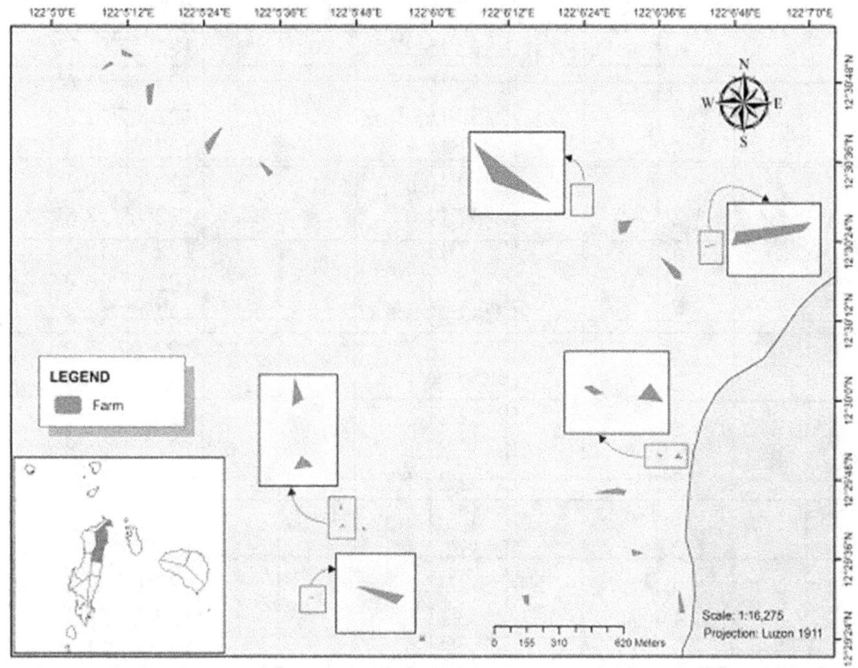

Figure 4. Tiger grass farms in Hinugusan, San Agustin, Romblon

Binongaan

Table 26. GPS coordinates of tiger grass farms in Binongaan

FARM	POINTS	NORTH LATITUDE	EAST LONGITUDE
Farm 1	1	12°31'48.0"	122°07'23.3"
	2	12°31'48.6"	122°07'22.5"
	3	12°31'50.1"	122°07'22.6"
	4	12°31'50.3"	122°07'23.0"
	5	12°31'49.3"	122°07'23.9"
Farm 2	1	12°32'04.0"	122°07'19.1"
	2	12°32'04.1"	122°07'18.5"
Farm 3	1	12°32'03.5"	122°07'15.5"
	2	12°32'03.1"	122°07'18.1"
	3	12°32'02.1"	122°07'17.9"
	4	12°32'02.4"	122°07'17.2"
Farm 4	1	12°32'31.8"	122°07'29.1"
	2	12°32'31.0"	122°07'29.2"
	3	12°32'30.6"	122°07'29.3"
	4	12°32'30.0"	122°07'29.6"
	5	12°32'28.9"	122°07'30.7"
Farm 5	1	12°32'25.1"	122°07'30.2"
	2	12°32'24.7"	122°07'30.4"
	3	12°32'24.5"	122°07'31.1"
	4	12°32'24.8"	122°07'31.7"
Farm 6	1	12°32'15.3"	122°07'34.3"
	2	12°32'16.0"	122°07'35.9"
	3	12°32'15.9"	122°07'34.3"
	4	12°32'15.1"	122°07'35.5"
Farm 7	1	12°32'15.8"	122°07'24.9"
	2	12°32'16.4"	122°07'23.8"
	3	12°32'15.4"	122°07'24.6"

FARM	POINTS	NORTH LATITUDE	EAST LONGITUDE
	4	12°32'15.2"	122°07'23.7"
Farm 8	1	12°32'16.4"	122°07'21.6"
	2	12°32'16.0"	122°07'22.2"
	3	12°32'14.8"	122°07'21.3"
Farm 9	1	12°32'10.5"	122°07'19.3"
	2	12°32'11.0"	122°07'18.6"
	3	12°32'11.7"	122°07'18.2"
	4	12°32'11.3"	122°07'16.9"
	5	12°32'09.6"	122°07'18.8"
Farm 10	1	12°32'26.7"	122°07'27.2"
	2	12°32'29.1"	122°07'25.5"
	3	12°32'27.3"	122°07'25.1"
	4	12°32'27.1"	122°07'26.8"
	5	12°32'28.4"	122°07'24.3"
Farm 11	1	12°32'21.9"	122°07'21.3"
	2	12°32'22.4"	122°07'21.6"
	3	12°32'22.6"	122°07'22.1"
	4	12°32'22.1"	122°07'22.3"
	5	12°32'21.0"	122°07'21.8"
Farm 12	1	12°32'38.1"	122°07'05.6"
	2	12°32'38.5"	122°07'07.2"
	3	12°32'38.2"	122°07'08.1"
	4	12°32'39.0"	122°07'09.1"
	5	12°32'39.9"	122°07'09.0"
	6	12°32'40.4"	122°07'08.5"
	7	12°32'40.7"	122°07'08.0"
	8	12°32'41.4"	122°07'06.7"
	9	12°32'42.0"	122°07'06.0"
	10	12°32'41.4"	122°07'05.1"
	11	12°32'41.6"	122°07'04.7"
	12	12°32'42.4"	122°07'04.7"
	13	12°32'42.6"	122°07'03.9"
	14	12°32'41.5"	122°07'03.7"
	15	12°32'40.4"	122°07'02.9"
	16	12°32'39.9"	122°07'02.6"
	17	12°32'39.7"	122°07'01.9"
	18	12°32'39.4"	122°07'01.6"
	19	12°32'39.1"	122°07'03.2"
Farm 13	1	12°32'26.9"	122°07'24.4"
	2	12°32'27.6"	122°07'22.9"
	3	12°32'27.4"	122°07'23.1"
	4	12°32'26.0"	122°07'23.4"
Farm 14	1	12°32'26.0"	122°07'26.9"
	2	12°32'25.8"	122°07'27.1"
Farm 15	1	12°32'01.2"	122°07'17.4"
	2	12°32'01.4"	122°07'17.9"
	3	12°32'02.0"	122°07'18.2"
Farm 16	1	12°32'04.8"	122°07'16.0"
	2	12°32'03.0"	122°07'14.1"
	3	12°32'04.8"	122°07'14.3"
	4	12°32'02.9"	122°07'14.0"
	5	12°32'03.7"	122°07'14.0"
Farm 17	1	12°32'04.9"	122°07'12.4"

FARM	POINTS	NORTH LATITUDE	EAST LONGITUDE
	2	12°32'03.6"	122°07'10.8"
	3	12°32'03.9"	122°07'10.6"
	4	12°32'06.5"	122°07'10.3"
	5	12°32'06.0"	122°07'11.3"

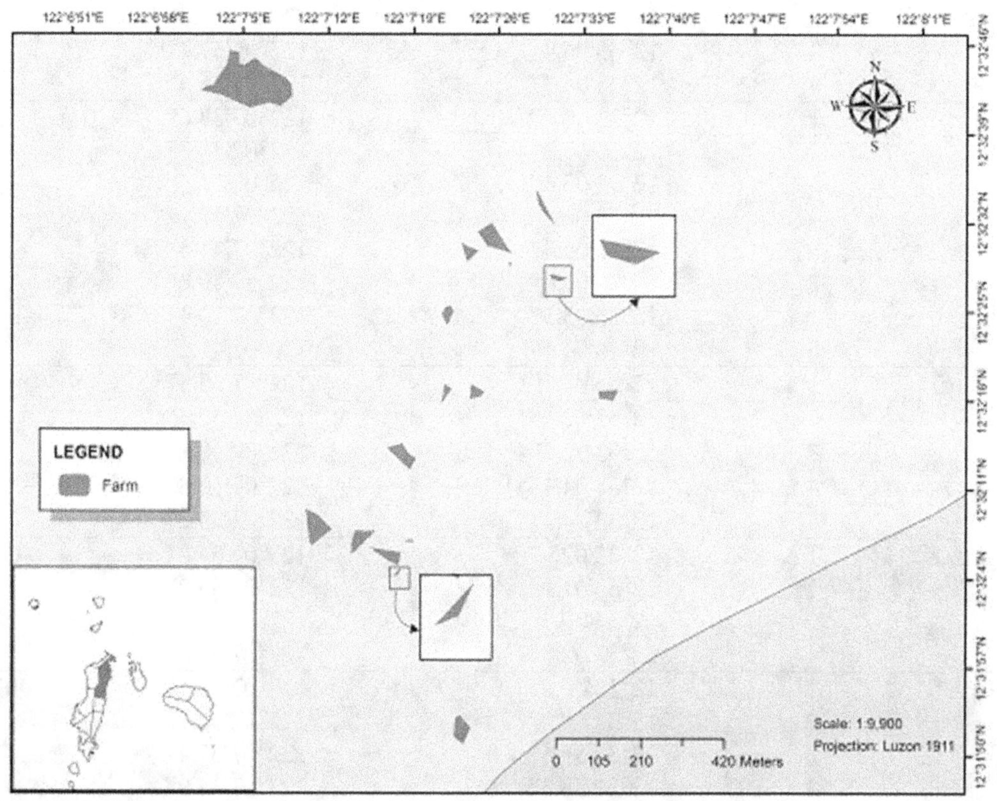

Figure 5. Tiger grass farms in Binongaan, San Agustin, Romblon

Doña Juana

Table 27. GPS coordinates of tiger grass farms in Doña Juana

FARM	POINTS	NORTH LATITUDE	EAST LONGITUDE
Farm1	1	12°33'02.3"	122°07'58.5"
	2	12°33'02.4"	122°07'58.4"
	3	12°33'01.9"	122°07'58.3"
	4	12°33'01.1"	122°07'57.2"
Farm 2	1	12°33'00.8"	122°07'40.0"
	2	12°32'59.0"	122°07'39.9"
	3	12°33'02.6"	122°07'45.3"
	4	12°32'59.7"	122°07'40.0"
	5	12°32'59.5"	122°07'42.8"
	6	12°33'02.2"	122°07'46.5"
	7	12°33'04.4"	122°07'43.8"
	8	12°33'04.6"	122°07'44.3"
Farm 3	1	12°33'07.1"	122°07'44.4"
	2	12°33'06.4"	122°07'43.6"
	3	12°33'07.1"	122°07'42.6"
	4	12°33'08.2"	122°07'43.8"
Farm 4	1	12°33'06.6"	122°07'41.9"

FARM	POINTS	NORTH LATITUDE	EAST LONGITUDE
	2	12°33'06.9"	122°07'40.9"
	3	12°33'07.7"	122°07'40.9"
	4	12°33'06.6"	122°07'38.3"
	5	12°33'05.7"	122°07'40.2"
	6	12°33'05.9"	122°07'42.0"
Farm 5	1	12°33'07.3"	122°07'34.5"
	2	12°33'07.0"	122°07'37.0"
	3	12°33'06.4"	122°07'37.4"
	4	12°33'04.3"	122°07'36.3"
Farm 6	1	12°33'03.9"	122°07'37.6"
	2	12°33'04.7"	122°07'36.7"
	3	12°33'05.6"	122°07'37.3"
	4	12°33'06.2"	122°07'39.1"
	5	12°33'05.4"	122°07'41.3"
	6	12°33'04.6"	122°07'40.8"
	7	12°33'03.5"	122°07'40.4"
	8	12°33'04.4"	122°07'41.2"
Farm 7	1	12°33'03.1"	122°07'35.3"
	2	12°33'02.6"	122°07'32.3"
	3	12°33'02.8"	122°07'30.8"
	4	12°33'04.1"	122°07'30.8"
	5	12°33'04.1"	122°07'33.2"
Farm 8	1	12°33'02.9"	122°07'34.8"
	2	12°33'01.0"	122°07'35.2"
	3	12°33'00.5"	122°07'34.7"
	4	12°33'02.8"	122°07'34.0"
Farm 9	1	12°33'00.2"	122°07'35.1"
	2	12°33'00.0"	122°07'38.1"
	3	12°32'55.6"	122°07'37.4"
	4	12°32'58.3"	122°07'34.9"
Farm 10	1	12°33'04.2"	122°07'31.9"
	2	12°33'04.2"	122°07'28.6"
Farm 11	1	12°32'59.1"	122°07'25.5"
	2	12°32'57.1"	122°07'26.2"
	3	12°32'58.9"	122°07'26.4"
	4	12°32'57.1"	122°07'25.1"
Farm 12	1	12°32'58.8"	122°07'24.1"
	2	12°32'54.6"	122°07'23.7"
	3	12°32'58.6"	122°07'21.5"
	4	12°32'56.6"	122°07'21.0"
Farm 13	1	12°33'21.0"	122°07'19.6"
	2	12°33'22.6"	122°07'20.7"
	3	12°33'24.4"	122°07'14.0"
Farm 14	1	12°33'26.6	122°07'11.4"
	2	12°33'28.4"	122°07'12.8"
Farm 15	1	12°33'18.6"	122°07'17.3"
	2	12°33'17.9"	122°07'18.8"
	3	12°33'17.0"	122°07'25.1"
	4	12°33'13.9"	122°07'23"
Farm 16	1	12°33'03.3"	122°07'50.3"
	2	12°33'03.1"	122°07'50.5"
	3	12°33'03.0"	122°07'50.4"
	4	12°33'03.1"	122°07'50.1"

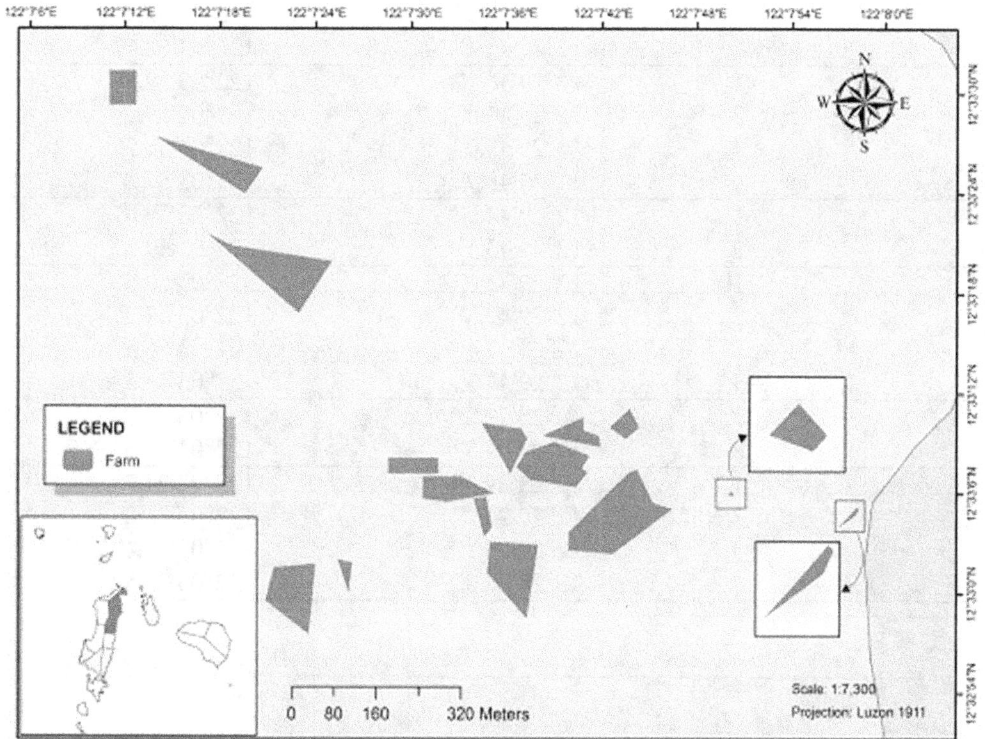

Figure 6. Tiger grass farms in Doña Juana, San Agustin, Romblon

Victoria

Table 28. GPS coordinates of tiger grass farms in Victoria.

FARM	POINTS	NORTH LATITUDE	EAST LONGITUDE
Farm 1	1	12°29'33.1"	122°03'46.3"
	2	12°29'34.5"	122°03'49.1"
	3	12°29'36.0"	122°03'49.4"
	4	12°29'34.4"	122°03'47.8"
	5	12°29'35.6"	122°03'49.0"
Farm 2	1	12°29'34.1"	122°03'49.5"
	2	12°29'34.8"	122°03'50.6"
	3	12°29'33.9"	122°03'50.0"
	4	12°29'33.6"	122°03'50.9"
	5	12°29'32.5"	122°03'49.7"
Farm 3	1	12°29'36.9"	122°03'49.3"
	2	12°29'38.9"	122°03'48.8"
	3	12°29'37.5"	122°03'48.4"
	4	12°29'38.6"	122°03'50.3"
	5	12°29'38.1"	122°03'50.4"
Farm 4	1	12°29'41.5"	122°03'53.9"
	2	12°29'40.6"	122°03'54.9"
	3	12°29'41.8"	122°03'54.5"
	4	12°29'39.6"	122°03'54.4"
Farm 5	1	12°29'39.0"	122°03'55.8"
	2	12°29'41.5"	122°03'57.0"
	3	12°29'40.4"	122°03'58.7"
	4	12°29'41.0"	122°03'56.1"
	5	12°29'41.2"	122°03'58.6"

FARM	POINTS	NORTH LATITUDE	EAST LONGITUDE
Farm 6	1	12°29'42.2"	122°04'02.5"
	2	12°29'43.4"	122°04'02.6"
Farm 7	1	12°29'46.4"	122°03'59.8"
	2	12°29'46.0"	122°03'59.7"
	3	12°29'45.0"	122°03'58.2"
	4	12°29'45.9"	122°03'57.2"
Farm 8	1	12°29'44.0"	122°03'59.0"
	2	12°29'43.3"	122°03'59.5"
Farm 9	1	12°29'43.0"	122°03'57.5"
	2	12°29'42.8"	122°03'57.2"
	3	12°29'43.4"	122°03'56.5"
	4	12°29'42.3"	122°03'55.5"
Farm 10	1	12°29'21.2"	122°03'43.0"
	2	12°29'19.5"	122°03'44.3"
Farm 11	1	12°29'27.3"	122°03'23.7"
	2	12°29'27.9"	122°03'21.8"
	3	12°29'29.1"	122°03'21.4"
	4	12°29'28.0"	122°03'23.5"

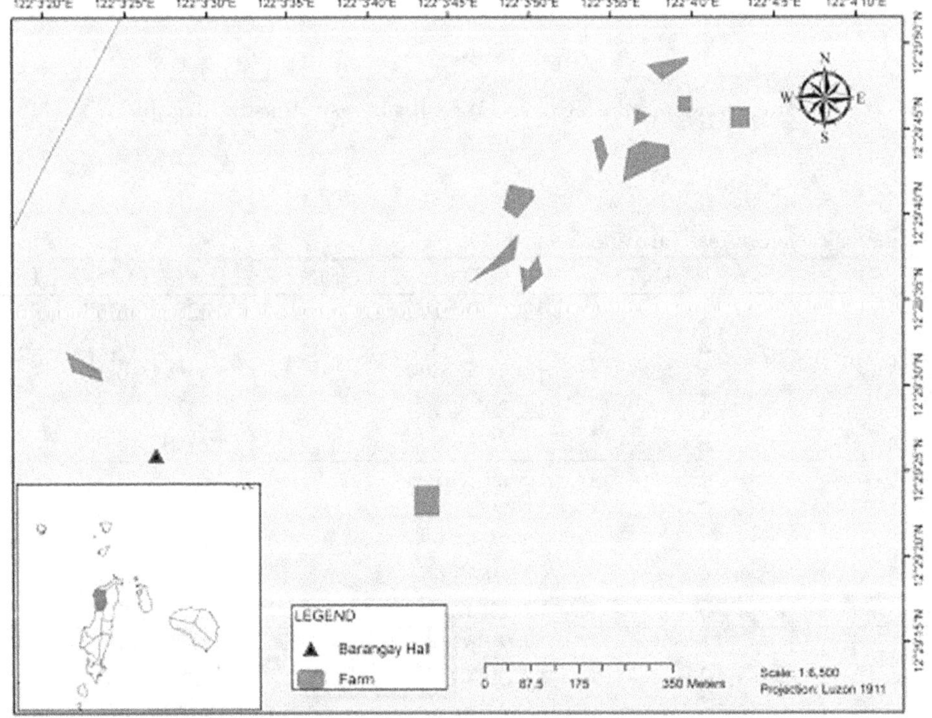

Figure 7. Tiger grass farms in Victoria, San Andres, Romblon

Pagsangahan

Table 29. GPS coordinates of tiger grass farms in Pagsangahan.

FARM	POINTS	NORTH LATITUDE	EAST LONGITUDE
Farm 1	1	12°35'25.3"	122°04'18.9"
	2	12°35'27.1"	122°04'17.6"
	3	12°35'26.6"	122°04'17.7"
	4	12°35'25.8"	122°04'17.8"
	5	12°35'27.4"	122°04'18.3"

FARM	POINTS	NORTH LATITUDE	EAST LONGITUDE
Farm 2	1	12°35'20.2"	122°04'19.2"
	2	12°35'21.7"	122°04'20.3"
	3	12°35'20.9"	122°04'18.8"
	4	12°35'20.6"	122°04'19.8"
Farm 3	1	12°35'12.8"	122°04'18.1"
	2	12°35'13.2"	122°04'18.7"
	3	12°35'13.0"	122°04'17.7"
	4	12°35'13.3"	122°04'17.9"
Farm 4	1	12°35'12.2"	122°04'20.3"
	2	12°35'11.5"	122°04'22.4"
	3	12°35'10.3"	122°04'21.0"
Farm 5	1	12°35'04.4"	122°04'31.5"
	2	12°35'03.7"	122°04'32.4"
Farm 6	1	12°35'04.3"	122°04'23.3"
	2	12°35'03.9"	122°04'22.9"
Farm 7	1	12°35'00.8"	122°04'13.5"
	2	12°34'58.9"	122°04'11.2"
	3	12°34'59.9"	122°04'11.7"
	4	12°34'59.5"	122°04'10.5"
	5	12°34'58.3"	122°04'11.9"
	6	12°35'01.6"	122°04'10.3"
Farm 8	1	12°34'56.3"	122°04'13.2"
	2	12°34'55.6"	122°04'13.3"
	3	12°34'56.1"	122°04'13.0"
	4	12°34'56.3"	122°04'14.1"
Farm 9	1	12°34'55.6"	122°04'16.7"
	2	12°34'55.7"	122°04'19.2"
	3	12°34'58.1"	122°04'18.8"
	4	12°34'56.3"	122°04'16.9"
	5	12°34'56.3"	122°04'19.5"
Farm 10	1	12°35'47.4"	122°04'24.9"
	2	12°35'48.1"	122°04'25.2"
	3	12°35'47.8"	122°04'25.3"
Farm 11	1	12°35'55.0"	122°35'55.6"
	2	12°35'55.6"	122°04'20.9"
	3	12°35'56.5"	122°04'20.8"
	4	12°35'56.6"	122°04'20.2"
Farm 12	1	12°35'38.4"	122°05'30.4"
	2	12°35'38.7"	122°05'31.6"
	3	12°35'38.9"	122°05'30.3"
	4	12°35'38.1"	122°05'31.4"
	5	12°35'39.1"	122°05'31.5"
Farm 13	1	12°35'42.9"	122°05'42.7"
	2	12°35'42.7"	122°05'43.8"
	3	12°35'43.0"	122°05'43.1"
	4	12°35'42.3"	122°05'44.1"
	5	12°35'42.1"	122°05'44.5"
Farm 14	1	12°35'38.7"	122°05'44.8"
	2	12°35'38.1"	122°05'44.9"
	3	12°35'37.0"	122°05'44.6"
Farm 15	1	12°35'37.6"	122°05'45.5"
Farm 16	1	12°35'42.7"	122°05'48.3"

FARM	POINTS	NORTH LATITUDE	EAST LONGITUDE
	2	12°35'42.5"	122°05'49.3"
	3	12°35'44.0"	122°05'49.0"
	4	12°35'42.3"	122°05'50.9"
Farm 17	1	12°35'37.7"	122°05'48.0"
	2	12°35'37.5"	122°05'48.5"
	3	12°35'37.6"	122°05'48.2"
	4	12°35'37.3"	122°05'48.3"
Farm 18	1	12°35'42.0"	122°05'39.7"
	2	12°35'43.1"	122°05'40.5"
	3	12°35'42.1"	122°05'39.5"
	4	12°35'42.5"	122°05'39.0"
Farm 19	1	12°35'40.5"	122°04'45.6"
	2	12°35'38.6"	122°04'44.4"
	3	12°35'40.0"	122°04'45.3"
	4	12°35'38.8"	122°04'45.2"
Farm 20	1	12°35'34.3"	122°04'43.2"
	2	12°35'32.4"	122°04'44.9"
	3	12°35'33.9"	122°04'43.5"
	4	12°35'33.1"	122°04'44.6"
Farm 21	1	12°36'26.9"	122°04'14.4"
	2	12°36'26.9"	122°04'16.0"
	3	12°36'27.0"	122°04'14.9"
	4	12°36'25.8"	122°04'15.4"
	5	12°36'25.5"	122°04'14.1"

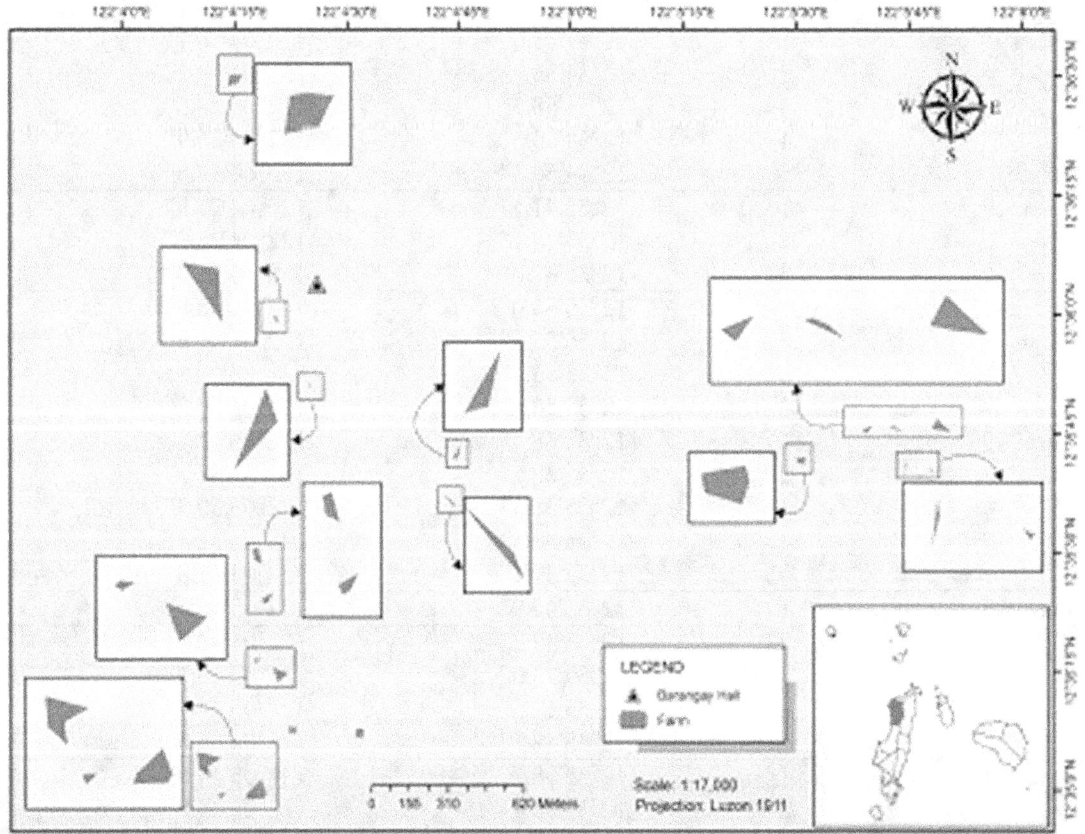

Figure 8. Tiger grass farms in Pagsangahan, Calatrava, Romblon

Mari-Sur

Table 30. GPS coordinates of tiger grass farms in Mari-Sur.

FARM	POINTS	NORTH LATITUDE	EAST LONGITUDE
Farm 1	1	12°31'29.3"	122°02'09.8"
	2	12°31'28.0"	122°02'10.0"
	3	12°31'28.4"	122°02'10.6"
	4	12°31'28.8"	122°02'09.4"
Farm 2	1	12°31'31.5"	122°04'04.7"
	2	12°31'31.0"	122°04'05.8"
Farm 3	1	12°31'31.6"	122°04'09.9"
	2	12°31'29.8"	122°04'10.1"
	3	12°31'31.0"	122°04'09.4"
	4	12°31'30.9"	122°04'08.9"
	5	12°31'31.8"	122°04'09.1"
Farm 4	1	12°31'21.5"	122°04'17.8"
	2	12°31'21.4"	122°04'19.5"
	3	12°31'21.1"	122°04'19.3"
Farm 5	1	12°30'55.5"	122°04'28.1"
	2	12°30'56.3"	122°04'29.5"
	3	12°30'55.7"	122°04'30.1"
	4	12°30'54.9"	122°04'29.2"
Farm 6	1	12°31'15.8"	122°03'02.7"
	2	12°31'18.2"	122°03'02.1"
Farm 7	1	12°31'25.6"	122°03'00.4"
	2	12°31'26.0"	122°03'01.1"
Farm 8	1	12°31'38.0"	122°03'12.8"
	2	12°31'38.3"	122°03'13.7"
Farm 9	1	12°32'02.8"	122°03'27.8"
	2	12°32'03.2"	122°03'29.1"
	3	12°32'02.1"	122°03'27.9"
Farm 10	1	12°32'04.0"	122°03'29.8"
	2	12°32'04.7"	122°03'29.3"
	3	12°32'04.2"	122°03'30.9"
Farm 11	1	12°30'16.2"	122°03'31.1"
	2	12°30'16.1"	122°03'32.6"
	3	12°30'17.0"	122°03'30.7"
	4	12°30'16.0"	122°03'32.3"
Farm 12	1	12°32'04.7"	122°04'06.0"
	2	12°32'04.5"	122°04'07.6"
	3	12°32'06.3"	122°04'07.3"
	4	12°32'05.2"	122°04'08.3"

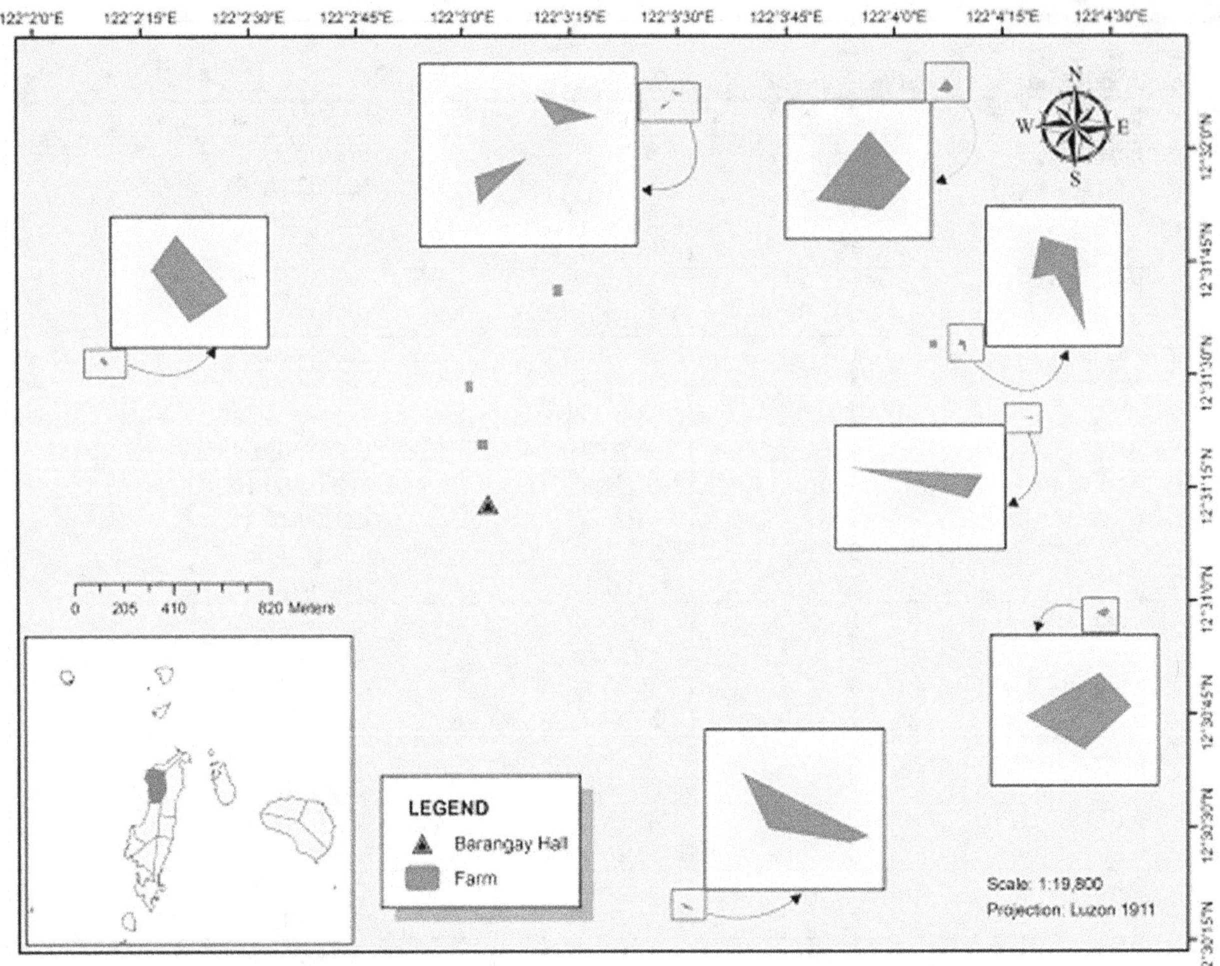

Figure 9. Tiger grass farms in Mari-Sur, San Andres, Romblon

Mari Norte

Although a baseline information for tiger grass industry already exists for Mari-Norte (Fetalvero & Faminial, 2010), two days were spent in getting only the coordinates for farm markers. This was due to the steep terrains in the area. With the help of these markers, which coincided with the farm area, farms could easily be identified.

Table 31. GPS coordinates (markers) of tiger grass farms in Mari-Norte.

FARM	MARKERS	NORTH LATITUDE	EAST LONGITUDE
Farm 1	1	12°34'43.4"	122°04'46.2"
	2	12°34'33.6"	122°04'49.3"
	3	12°34'31.7"	122°04'47.6"
	4	12°34'30.7"	122°04'47.8"
Farm 2	1	12°34'30.5"	122°04'48.1"
	2	12°34'29.0"	122°04'49.0"
	3	12°34'26.5"	122°04'49.3"
	4	12°34'20.6"	122°04'50.8"
	5	12°34'20.3"	122°04'49.6"
Farm 3	1	12°34'20.8"	122°04'48.1"
	2	12°34'19.7"	122°04'47.2"
	3	12°34'18.0"	122°04'46.6"
	4	12°34'16.8"	122°04'47.0"
	5	12°34'15.3"	122°04'46.1"

FARM	MARKERS	NORTH LATITUDE	EAST LONGITUDE
Farm 4	1	12°34'09.4"	122°04'46.6"
	2	12°34'10.4"	122°04'41.4"
	3	12°34'09.7"	122°04'34.5"
Farm 5	1	12°34'03.8"	122°04'30.0"
	2	12°34'00.3"	122°04'28.5"
Farm 6	1	12°34'13.3"	122°04'47.2"
	2	12°34'13.0"	122°04'47.1"
Farm 7	1	12°34'15.0"	122°04'54.6"
	2	12°34'14.9"	122°04'54.2"
	3	12°34'16.2"	122°04'53.2"
Farm 8	1	12°34'15.9"	122°04'52.9"
	2	12°34'15.7"	122°04'49.2"
Farm 9	1	12°34'19.4"	122°05'09.2"
	2	12°34'19.7"	122°05'10.1"
Farm 10	1	12°34'20.7"	122°05'11.3"
Farm 11	1	12°33'54.9"	122°04'26.5"
	2	12°33'53.5"	122°04'25.2"
	3	12°33'51.6"	122°04'23.1"
	4	12°33'46.3"	122°04'18.3"
Farm 12	1	12°33'40.2"	122°04'10.8"
	2	12°33'36.1"	122°04'06.2"
Farm 13	1	12°33'33.7"	122°04'03.4"
Farm 14	1	12°33'25.9"	122°03'59.2"
Farm 15	1	12°33'59.6"	122°03'51.1"
Farm 16	1	12°32'39.6"	122°03'57"
Farm 17	1	12°32'38.9"	122°04'16.4"
	2	12°32'37.6"	122°04'16.0"
	3	12°32'36.6"	122°04'15.6"
	4	12°32'37.1"	122°04'14.4"
	5	12°32'37.7"	122°04'10.7"
	6	12°32'38.2"	122°04'09.7"
Farm 18	1	12°33'44.0"	122°03'56.2"
	2	12°33'42.2"	122°04'02.3"
	3	12°33'40.9"	122°04'04.0"
	4	12°33'40.6"	122°04'04.8"
Farm 19	1	12°35'55.4"	122°03'47.2"
	2	12°33'51.8"	122°03'48.5"
	3	12°33'51.2"	122°03'49.0"
	4	12°33'46.6"	122°03'54.4"
Farm 20	1	12°33'34.1"	122°03'54.5"
	2	12°33'34.6"	122°03'55.0"
Farm 21	1	12°33'34.6"	122°03'46.3"
Farm 22	1	12°33'37.5"	122°03'40.9"
Farm 23	1	12°33'38.5"	122°03'38.0"
Farm 24	1	12°33'37.8"	122°03'30.6"
	2	12°33'41.3"	122°03'29.7"
Farm 25	1	12°33'49.7"	122°03'14.3"

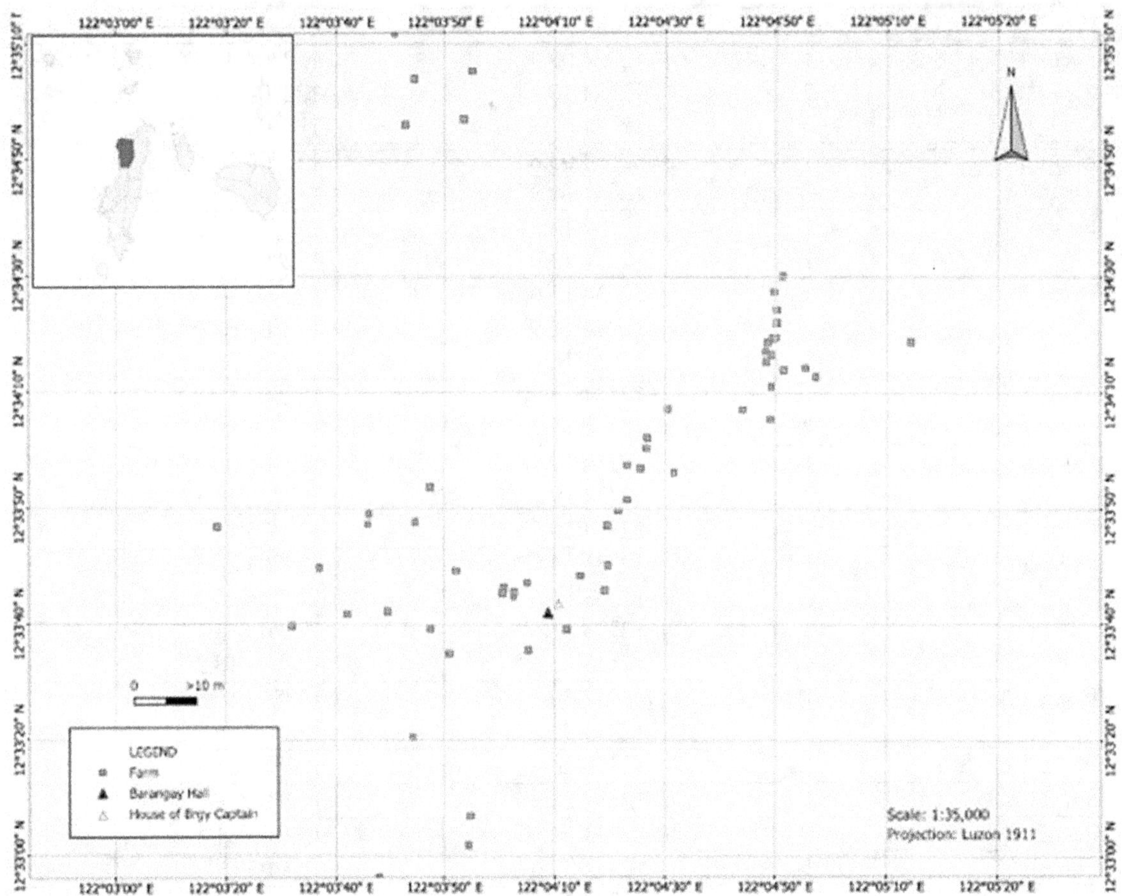

Figure 10. Markers for tiger grass farms in Mari-Norte, San Andres, Romblon

9 CONCLUSIONS AND RECOMMENDATIONS

Based from the findings of this study was the following conclusions were formulated:

1. There is a strong literature base and scientific evidence supporting the potential and significance of tiger grass, *Thysanolaena maxima* (Roxb.) O. Ktze. as a multi-purpose crop. Its economic importance is foremost advanced. This grass holds a promise in phytoremediation, bioengineering, medicine, agriculture, climate change adaptation, and more.

2. There was a narrow gap in terms of percentage of males and females involved in tiger grass industry. Most of them were married with an average of four dependents. Most were 46 years old and their nature of work related to the tiger grass industry was either a farmer or farmer-processor. Very few of them concentrated in tiger grass processing alone.

3. The socio-economic condition of the respondents was characterized by low educational attainment and low estimated annual income indicating that they belong to the social poor, if poverty is measured by these two indicators. Other income sources were limited to seasonal agricultural activities like copra production and palay farming. There was a growing interest in social participation triggered by DSWD's 4Ps program.

4. For the last five years, the increasing number of people interested in tiger grass farming was observed. These farms averaging 1 hectare per farmer were commonly managed by the tenants or owned by the respondents. They were estimated to be 3km away from their home. Under crops like coconut and palay were planted alongside tiger grasses.

5. The tiger grass farming industry required minimal farm inputs. Those with vast plantations spent around P5,000 on the average which they usually paid to farm workers who were usually relatives and family members. The common daily wage was P150 per day. Indigenous farm implements like bolo, tara- tara and tagad were commonly used in farm preparation and planting.

6. The annual production volume of flowers ranged from 120 to 300 bundles per farmer where bundle cost ranged from P25 during peak months and P52 during off- peak months yielding an average income estimate of P3,750. However, a large percentage of the respondents refused to declare their income with the hesitation that they would be made to pay taxes or they would be removed from the list of DSWD's indigents. Should they cooperated in the survey, the income estimate could have been higher.

7. The average length of processors engagement into soft broom industry was 10 years. Regular and jumbo soft brooms were sold from P25 to P60 respectively. The average capital needed in soft broom making was P3,000. This was self-funded. Soft broom making can assure every processor an annual income estimate of P17,500. But this result was again another understatement because there were a large percentage of the respondents who refused to give their actual income afraid that they may be taxed or removed from the list of the indigents.

8. Soft brooms had already established a number of markets within and outside the province but most of the products were concentrated in Dona Juana. Marketing practices were of the traditional type but there were developments as to product packaging. Some processors were already aware of competition and were observing quality control, branding (Tablas premium) and product improvement. However, the 'Made in Baguio' labels were still prevalent.

9. The most common problems faced by the farmers were infestation of pests and animals, theft and climate change while those of the processors were the tiresome processing activity, lack of capital and mold attack during rainy days which were brought about by climatic changes.

10. San Agustin has the largest plantation of tiger grass yet in the province as compared to those in San Andres and Calatrava.

Recommendations

Based from findings and conclusions of the study, the following have been recommended:

1. The identified resource maps in this study must be integrated into the Community-Based Resource Management map of the barangay, town and province for proper monitoring of the community's resources.

2. There is a need to fast track the formulation of a Tiger Grass R&D program that will set the roadmap towards consideration of the commodity as among the provincial, regional and national priorities. The RSU-Research, Extension and Production Unit can form a committee to organize the comprehensive R&D plan for tiger grasses particularly its industrial, medicinal, agricultural and environmental significance.

3. The Extension Unit of the College of Business and Accountancy of Romblon State University can initiate programs that will prepare these farmers and processors for a cooperative organization. Aside from pre-membership seminars, business –related knowledge on some aspects of the industry can also be extended. It can also package an information and education campaign on the potentials of tiger grass so that the community people will be made aware of its importance.

4. The developed prototype machine, the tiger grass pollen remover with woodworking tool, of the College of Engineering and Technology of Romblon State University can be tested in these barangays to unburden the farmers and processors of the physically taxing manual work. Likewise, a technology that can speed up the soft broom processing may also be explored upon.

5. The College of Business and Accountancy of Romblon State University in coordination with DTI, NEDA, DOST and DA can pilot the previously proposed "Entrepreneurial Camp" as an integration to the BSBA curriculum. This proposed encampment will be a two-month immersion activity of the business students in a potential community with economic activity to boast like tiger grass industry of these barangays. During the encampment, the students will be taught and guided on the preparation of a project proposal and the best proposal could be packaged for possible funding. This activity is expected to broaden the social concept of the students and learn on hand what community development is all about.

6. On gender and development, a program can be packaged to empower women tiger grass farmers and processors by the GAD coordinators in the barangay or other agencies concerned.

7. On further studies, the following may be conducted or initiated:
    a. Profile of tiger grass industries in barangay Lusong and Salingsing in San Agustin;
    b. An *in situ* study of cultural practices on tiger grass farming focusing on plantation intervals and fertilizer applications;
    c. Cost analysis and market channel studies of soft brooms;
    d. Impact of climate change on the tiger grass industry; and
    e. Validation and trial studies on the reported uses of this plant as reported in literatures.

LITERATURES CITED

[1] Abad, F. (2008). Tiger grass flower pollen remover. *Philippine Journal of Crop Science*, 131.

[2] Areekul S., Sinchaisri, P. and Tigvatananon S. (1988). Effect of Thai plant extracts on the oriental fruit fly III. attractancy test. *Kasetsart Journal: Natural Sciences*, 22 (2) : 160-164.

[3] Aryal, K.P., Berg, A. and Ogie, B. (2009). Uncultivated plants and livelihood support: A case study from the Chepang people of Nepal. *Ethnobotany Research and Applications*, 17 : 409-422.

[4] Baldino, T. (2002, January-February). Growing tiger grass under Benguet pine stand. *Canopy International*, pp. 2, 11.

[5] Bhardwaj, S. and Gakhar, S.K. (2008). Ethnobotanicals used by tribals of Mizoram for furniture and household equipments. *Indian Journal of Traditional Knowledge*, 7 (1): 134-137.

[6] Bhuchar, S. (2008). Broom grass: A multipurpose plant with erosion control potential. *HMCAT Newsletter*, pp. 16-18.

[7] Bisht, N.S. and Ahlawat, S.P. (1998). *Broom Grass.* Arunachal Pradesh, Itanagar, India: State Forest Research Institute.

[8] Broemme, K. and Stolpe H. (2011). *Mining and Environment in Vietnam, Research Work of the Research Association Mining and Environment (RAME) Status Report.* Vietnam: RAME.

[9] Brunings, A.M., Datnoff, L.E., Palmateer, A.J., Locke, J.C. and Krause, C.R. (2009). *Exserohilum* leaf spot on tiger grass. *Plant Health Progress*, Online.

[10] Caringal, A.M. and Bañados, H.G. Jr. (2008). Business ecologyof non-timber forest products (NTFPs) utilized under subsistence and small-cash household economy: a survey of indigenous mountain products from southern Batangas, Philippines. *Journal of Nature Studies*, 7 (1): 183-191.

[11] Costales, E. (1985). Determination and evaluation of some emergency measures for quick rehabilitation of newly burned areas in the pine forest watershed. *Proceedings of the Seminar on Watershed Research and Management Practices, ASEAN-US Watershed Project* (pp. 113-132). Serdang, Selangor, Malaysia: College, Laguna, Philippines.

[12] Costales, E.F. Jr. and Costales A.B. (1985). Effects of plant combination on the protection/stabilization of mined waste areas. *Sylvatrop*, 10 (2) : 187-202.

[13] Dogan, Y., Nedelcheva, A.M. and Yarci, C. (2008). Plant taxa used as brooms in several Southeast European and West Asean Countries. *Nat. Croat.*, 17 (3): 193-206.

[14] *Dok khaem.* (n.d.). Retrieved March 10, 2012, from Dok khaem: http://www.tabi.la/lao-ntfpwiki/index.php/Dok_khaem

[15] ECS. (2008). *Treatment of land Slide and erosion Control Work under "TDET" Project, South Sikkim 2004-05 to 2007-08.* South Sikkim, India: Envis Centre Sikkim.

[16] Fetalvero, E.G., Faminial, T.F. and Sespeñe, J.S. (2011). Tiger grass industry in Marigondon Norte, San Andres, Romblon: Implications for research and development. *Travesia* , 1(1):81-95.

[17] Fetalvero, E.G. and Faminial, T.T. (2010). *Tiger Grass Industry in Marigondon Norte, San Andres, Romblon: Implications for Research and Development.* Technical Report submitted to DTI-Romblon

[18] Himmelsbach, W., Tagle, M.A.G., Fuldner, K. Hoefle, H.H. and Htun, W. (2006). Food plants of captive elephants in the Okkan Reserved Forest, Myanmar (Burma), Southeast Asia. *Ecotropica* , 12:15-26.

[19] Huque, K.S., Rahman and M.M., Jalil, M.A. (2001). Nutritive value of major feed ingredients, usually browsed and their responses to gayals *(Bos frontalis)* in the hill tract area. *Pakistan Journal of Biological Sciences* , 4 (12) : 1559-1561.

[20] Hynniewta, S.R. and Kumar Y. (2008). Herbal remedies among the Khasi traditional healers and village folks in Meghalaya. *Indian Journal of Traditional Knowledge* , 7(4):581-586.

[21] Jagadev, R.S. and Patnaik, A.K. (1994). Genetic variability and character association in broomgrass *(Thysanolaena maxima Roxb.). The Indian Journal of Genetics and Plant Breeding* , 54 (3) : 300-303.

[22] Jana, S.K. and Chauhan, A.S. (2000). Ethnobotanical studies on Lepchas of Dzongu, North Sikkim. *Ann For* , 8(1):131.

[23] Joshil, N.P. and Singh, S.B. (1989). Availability and use of shrubs and tree fodders in Nepal. *Proceedings of a workshop,* (pp. 211-220). Densapar, Indonesia.

[24] Kafle, G. (2005). *Evaluation of effectiveness of root and foliage system of grasses used in soil conservation.* Kathmandu, Nepal: Unpublished Thesis at Tribhuvan University.

[25] Khadka, C. (2011). *Impacts of climate change on production of cash crops in Annapurna Conservation Area: A case study from Lwang Ghalel Village Development Committee, Kaski District.* Kathmandu, Nepal: Unpublished Thesis at Tribhuvan University.

[26] Kharkongor, P. and Joseph J. (1981). Folklore medicobotany of rural Khasi and Jaintia tribes in Meghalaya. In E. B. Jain, *Glimpses of Indian Ethnobotany* (p. 115). New Delhi, India: Oxford and IBH.

[27] Khisa, K. (2001). Contour hedgerow inter-cropping agroforestry technology for degraded hillside farms at Chittagong Hill Tracts. *National Wrkshop on Agroforestry Research* (pp. 179-184). Khagrachari: Chittagong Hill Tracts Development Board.

[28] Khisa, S.K., Alam, M.K. and Siddiqi, N.A. (1999). Broom grass *(Thysanolaena maxima)* hedges: a bioengineering device for soil erosion control and slope stabilization. *First Asia-Pacific Conference on Ground and Water Bioengineering for Soil Erosion Control and Slope Stabilization,* (pp. 143-149). Manila.

[29] Lasco, R.D., Rangasa, M., Pulhin, F.B. and Delfino R.J. (2008). *The Role of Local Government Units in Mainstreaming Climate Change Adaptation in the Philippines. Center for Initiatives and Research on Climate Change Adaptation.*

[30] Livestalk. (2011, October). Livestock keepers dependent on natural rangeland.

[31] Lyngdoh, E.K. and Baishya, R. (2010). *People's perception on climate change: A case study from Meghalaya.* New Delhi, India: LEAD India.

[32] Mahato, R.B. and Chaudhary R.P. (2005). Ethnomedicinal study and antibacterial activities of selected plants of Palpa District, Nepal. *Scientific World*, 3(3):26-31.

[33] Maity, D., Pradhan, N., and Chauhan, A.J. (2004). Folk uses of some medicinal plants from North Sikkim. *Indian Journal of Traditional Knowledge* , 3 (1):66-71.

[34] Maki, F, Mamoru K., Thein, H.M. and Minn, Y. (2007). Recovery process of fallow vegetation in the traditional Karen Swidden Cultivation System in the Bago Mountain Range, Myanmar. *Southeast Asian Studies* , 45(3): 303-316.

[35] Malla, B. and Chhetri, R.B. (2009). Indigenous knowledge on ethnobotanical plants of Kavrepalanchowk District. *Kathmandu University Journal of Science, Engineering and Technology* , 5 (2): 96-109.

[36] Marrier, T. E. (2011, December). The Aeta-Pinatubo loop. *Communicative and Integrative Biology 4:6* , pp. 788-790.

[37] Mathema, P. and Joshi, J. (2010). Assessment of small-scale landslide treatment in Nepal. *Banko Janakari* , 20 (91):3-8.

[38] Mau, R.F.L. and Matin, J.L. (2007, April). *Bacteria dorsalis (Hendel).* Retrieved March 10, 2012, from Crop Knowledge Master: http://www.extento.hawaii.edu/Kbase/Crop/Type

[39] Namsa, N.D., Mandal, M., Tangjang, S. and Mandal S. (2011). Ethnobotany of the Monpa ethnic group at Arunachal Pradesh, India. *Journal of Ethnobiology and Ethnomedicine* , 7:31.

[40] NCVST. (2009). *Vulnerability in the eyes of the vulnerable.* Nepal Climate Vulnerability Study Team.

[41] Nicholson, K., Ketphanh, S. and Sengdala, K. (2008). *Mission Report: Review of Experience in the Marketing, production, Harvesting and Management of Agro-biodiversity and NTFP products for the agro-biodiversity initiative.* Laos.

[42] Orallo. (2007). Tiger grass farming and broom making in Bagulin, La Union, Philippines. *PCARRD Highlights* , 99-100.

[43] OTOP. (n.d.). *About San Andres.* Retrieved February 1, 2009, from OTOP Philippines: http://www.otopphilippines.gov.ph/microsite.aspx?rid=5&provid=67&prodid-157&sec=2

[44] OTOP. (n.d.). *One Cluster One Vision.* Retrieved February 1, 209, from DILG: http://www.dilg.gov.ph/Region4B/bestpractices.htm

[45] OTOP. (n.d.). *Tiger Grass of San Andres.* Retrieved February 1,2009, from OTOP Philippines: http://otopphilippines.gov.ph/microsite.aspx?rid=5&provid=67&townid=1277&prodid=157

[46] OTOP. (n.d.). *Tiger Grass Processing.* Retrieved February 1, 2009, from DOST: http://mis.dost.gov.ph/region4/webapps/ProjMonitor/details.php?id=72

[47] OTOP. (n.d). *Tiger Grass Producers in San Andres.* Retrieved February 1, 2009, from OTOP Philippines: http://www.otopphilippines.gov.ph?sme.aspx?smeid=29&rid=5&provid=67&proid=157

[48] Pal D. and Jain S.K. (1998). *Tribal Medicine (Naya, Prokash, Kolkata).*

[49] Pal, D. (1991). Grasses of Bangladesh: an overview based on annotated specimens at the Central National Herbarium at Calcutta. *Proceedings of the International Botanical Conference* (p. 5). Dhaka, Bangladesh: BBS, Dhaka, Bangladesh.

[50] Palma, N.A. (1993). *Sucker and Panicle Production of Tiger Grass in Plantations.* Retrieved February 1,2009, from PCARRD:http://www.pcarrd.dost.gov.ph/consortia/nomcarrd/researches/1993/1993_r12.htm

[51] Pandit, B.H., Albano, A. and Kumar, C. (2008). Improving forest benefits for the poor: Learning from community-based forest enterprises in Nepal. Bogor, Indonesia: Center for international Forestry Research.

[52] Quiachon, L. and Tagra, M. (2002). Tiger grass *(Thysanolaena maxima)* production livelihood project. *PCARRD Highlights* , 113-114.

[53] Quiachon, L. (2002). Effect of spacing on the panicle production of tiger grass *(Thysanolaena maxima)* at Camp 7, Minglanilla, Cebu, Philippines. *Ecosystems Research Digest* , 12 (2) : 68-77.

[54] Rai, M. B. (2003). Medicinal plants of Tehrathum District, Eastern Nepal. *Our Nature* , 1:42-48.

[55] Ramm Botanicals. (2009, May). *Colours of paradise.* Retrieved March 10, 2012, from Ramm Botanicals: http://www.ramm.com.au

[56] Razzaque, M.A. and Khan, M.S. (1978). Insulation boards from five grass species of the Chittagong region in Bangladesh. *Bano Biggyan Patrika* , 7 (1-2): 30-37.

[57] RFRI. (2008). Improvement of degraded shifting cultivation through the introduction of *Thysanolaena maxima* along with *Cajanas cajan* as N_2 fixing plant. In R. Jorhat, *Annual Report 2007-2008* (p. 16 pages). Rain Forest Research Institute.

[58] Rohilla, P.P. and Bujarbaruah, K.M. (2000). Effect of feeding broom grass *(Thysanolaena maxima)* to rabbits. *Indian Journal of Animal Nutrition* , 17 (1):87-89.

[59] Ronya, L. (1998). Foreword. In N. B. Ahlawat, *Broom Grass.* Arunachal Pradesh, Itanagar, India: State Forest Research Institute.

[60] Roothaert, R.L., Binh, L.H., Magboo, E., Yen, V.H. and Saguinhon, J. (2005). Participatory forage technology development in Southeast Asia. *Proceedings of the 12th Annual Conference of Ethiopia Society of Animal Production.* Addis Ababa: ESAP.

[61] Rotkittikhun, P., Chaiyarat, R., Kruatrachue, M., Pokethitiyook, P. and Baker, A.J.M. (2007). Growth and lead accumulation by the grasses *Vetiveria zizanioides* and *Thysanolaena maxima* in lead-contaminated soil amended with pig manure and fertilizer: A glasshouse study. *Chemosphere* , 45-53.

[62] Saikia, D.C., Goswami, T., Chaliha, B.P. (1992). Paper from *Thysanolaena maxima. Bioresource Technology* , 245-248.

[63] Sarmah, R. and Arunachalam, A. (2011). Contribution of non-timber forest products (NTFPs) to livelihood economy of the people living in forest fringes in Changlang District of Arunachal Pradesh, India. *Indian Journal of Fundamental and Applied Life Sciences* , 157-169.

[64] Saxena, K.G. and Ramakrishnan, P.S. (1983). Growth and allocation strategies of some perennial weeds of slash and burn agriculture (jhum) in northeastern India. *Canadian Journal of Botany* , 61 (4): 1300-1306.

[65] Sengupta, S., Gopal, B. and Das, S.N. (2004). Effect of nutrient supply and water depth on nutrient uptake by two wetland plants. *Bulletin of the National Institute of Ecology* , 14:55-60.

[66] Servañez B.F and Servañez, M.V. (n.d.). *The Tiger Grass Industry in Romblon – Rising from the Grassroots.* Document available at the DOST-PSTC Romblon Office.

[67] Servañez, M. (n.d.). *A Project Proposal on Techno-Demo cum Forum on Tiger Grass Technologies in San Andres Romblon.* Document available at the DOST-PSTC Romblon Office.

[68] Shankar, U., Lama, S.D. and Bawa, K.S. (2001). Ecology and economics of domestication of non-timber forest products: an illustration of broomgrass in Darjeeling Himalaya. *Journal of Tropical Forest Science* , 13 (1) : 171-191.

[69] Sharma, C. (2004). Indigenous soil erosion control and slope stabilization techniques in the hills and mountains of Nepal. *Ground and Water Bioengineering for Erosion Control and Slope Stabilization* , 207-215.

[70] Shrestha, P. (1985). Research note: Contribution to the ethnobiology of the Palpa area. *CNAS Journal* , 63-74.

[71] Singh, H.B., Prasad, P. and Rai, L.K. (2002). Folk medicinal plants in the Sikkim Himalayas of India. *Asian Folklore Studies* , 61:295-310.

[72] Srijumpa, N. and Seehawong, S. (2002). Some weedy grasses as substrates for mushroom cultivation. *Agriculture Science Journal* , 33 (6) : 297-306.

[73] Srijumpa, N. (2002). Use of some grasses as the substrates for *Pleurotus sp.* cultivation. *Thai Agricultural Research Manual* , 20 (1) : 3-8.

[74] Stapleton, C. (1989). *Chirang Hill Irrigation Project Damphu: Identification of Bamboos and Potential for Incorporating the Planting of Bamboos into Existing and Possible Activities.* Taba: Forestry Research Division.

[75] Subba, D. B., Thorne, P.J., Omed, H.M. and Sinclair, F.L. (2004). Investigating the biological interpretation of *adilopan* (appetite satisfaction), a term used by Nepalese hill farmers to evaluate fodder quality. *Proceedings of the Annual Meeting of the British Society for Animal Science* (p. 63). University of York, UK: British Society for Animal Science.

[76] Subba, D., Thorne, P. and Sinclair, F.L. (2002). Using local knowledge as a basis for planning ruminant diets in the mid hills of Nepal. *International Conference on Responding to the Increasing Global Demand for Animal Products*, (pp. 238-239). Merida, Mexico.

[77] Sudshishri, S., Dass, A. and Lenka N.K. (2008). Efficacy of vegetative barriers for rehabilitation of degraded hill slopes in eastern India. *Soil and Tillage Research* , 99(1):98-107.

[78] Tang, Y., Cao, M. and Fu, X. (2006). Soil seedbank in a dipterocarp rain forest in Xishuangbanna, Southwest China. *Biotropica* , 38 (3): 328-333.

[79] Tiwari, B. K. (2001). Domestication of three non-traditional species by shifting cultivars in India. *Workshop on Shifting Cultivation: Towards Sustainability and Resource Conservation in Asia, Cavite (Philippines)* (pp. 98-104). IRR Silang, Cavite, Phils.

[80] Tuddao, V. Jr. and Evasco, F. (1996). Increased panicle production of tiger grass *(Thysanolaena maxima)* through effective planting stock production and plantation establishment technique. *PCARRD Highlights* , 27.

[81] Viernes, R. (n.d.). *A Project Proposal on the Utilization of Tiger Grass Materials into Soft Brooms by Tiger Grass Broom Weavers of Mari-Norte, San Andres, Romblon.* Document available at the Office of the Municipal Agriculturist, San Andres, Romblon.

[82] Wang, Y.Z., Wong, M.K.M. and Hyde, K.D. (2000). Ommatomyces, with one new species and one new combination. *Fungal Diversity* , 4:125-131.

[83] Weed Watch Australia. (2011, November). *Tiger Grass* . Queensland, Australia: Technigro Australia.

[84] Wetterland, O., Zingerli, C. and Sorg J.P. (2004). Non-timber forest products in Nam Dong District, Central Vietnam: Ecological and economic prospects. *Schweiz Z. Fortwes* , 155 (2): 45-52.

[85] Yimyong, S., Sangwantanaroj, U. and Punnapayak, H. (2005). Use of weeds for the production of cellulase and ethanol. *1st International Conference on Fermentation Technology for Value Added Agricultural Products*, (pp. P-NF21). Khon Kaen, Thailand .

[86] Zhang, L.L.G., Liu, C.Q., Liu, H., Xiang, M. and Wei, X. (2011). Antimony mobility and transport in mine dumps from the Dachang multi-metalliferous mine area, Guangxi, China. *Proceedings of the 2nd International Workshop on Antimony in the Environment* (p. T22). Jena, Germany: University of Geneva and Friedrich-Schiller University.

[87] ____________(n.d.) . *A Training Design on the Operation and Maintenance of Tiger Grass Kiln Dryer and Broom Handle Making* Equipmen in Doña Juana, San Agustin, Romblon. Document available at the DOST-PSTC Romblon Office.

[88] ______________(1992). *Propagation, management and harvesting of Tiger grass in the highlands.* DENR-CAR, Technology Transfer Series, Vol. 2 (2). Retrieved on February 20, 2010 from http://maidon.pcarrd.dost.gov.ph/cin/afin/propagation-management-and-harvesting-of-tiger-grass-in-the-highlands.htm

[89] _____________(n.d.). *Quick Facts: The Tiger Grass Broom Industry in the Province of Romblon.* Document available at the DOST-PSTC Romblon Office.

[90] _____________ (2004). *Tiger Grass Industry Porfile at Doña Juana, San Agustin, Romblon.* Document available at the DOST-PSTC Romblon.

[91] _____________(n.d.). *Tiger Grass Processing: A Project Proposal of theTiger Grass Farmers and Processors Association at Doña Juana, San Agustin, Romblon.* Document available at the DOST-PSTC Romblon Office.

Appendix A

List of Respondents, Farm Area and Production Volume (brooms)

	VILLAGE	FARM AREA	PRODUCTION VOLUME (brooms)
HINUGUSAN			
	FARMERS		
1	Doroy, Ricky	2	.
2	Doroy, Ronel	2	.
3	Flaviano, Levy	1	.
4	Magramo, Nilda	0.5	.
5	Manong, Aladino	2	.
6	Manzo, Diosdado	2	.
7	Manzo, Eddie	3	.
8	Manzo, Edwin	1	.
9	Manzo, Ermie	2	.
10	Manzo, Joel	1.5	.
11	Miñon, Reynante	1	.
12	Miñon, Roger	1.5	.
13	Morada, Elma	4	.
14	Morada, Teresita	5	.
15	Regala, Artemio	0.6	.
16	Salvador, Armando	3	.
	BOTH FARMERS & PROCESSORS		
1	Manzo, Metaly	**0.5**	.
2	Orencio, Virgilio	1	50
BINONGAAN			
	FARMERS		
1	Dela Vega, Maribeth	1	.
2	Domingo, Tessie	0.25	.
3	Enad, Ruel	1	.
4	Flaviano, Cenon	1	.
5	Flaviano, Diosito	0.5	.
6	Flaviano, Rolando	0.5	.
7	Mago, Sixto	1	.
8	Manalon, Dante	3	.
9	Manalon, Renelyn	0.5	.
10	Manayon, Simplecia	0.5	.
11	Mangaya, Zander	1	.
12	Manlolo, Balbino	4	.
13	Manlolo, Melvin	0.5	.

	VILLAGE	FARM AREA	PRODUCTION VOLUME (brooms)
14	Mariño, Josita	2	.
15	Mariño, Paulino	3	.
16	Menes, Germelyn	1.5	.
17	Menes, Weneng	1	.
18	Mercano, Edgar	0.5	.
19	Mirallanes, Dalmacio	1	.
20	Miralles, Rolly	7	.
21	Montera, Modesto/Elena	2	.
22	Moreno, Aris	1	.
23	Moreno, Byron	1	.
24	Mortera, Alexander	4.5	.
25	Roldan, Joseph	1	.
26	Serabia, Edgar	1.5	.
27	Villacruses, Maly	0.5	.
28	Villaluren, Lorna	0.75	.
	FARMERS & PROCESSORS		
1	Gabo, Geraline	1	800
2	Galang, Celesty	0.5	200
3	Galindez, Reynaldo	1	300
4	Gonzales, Juanito	2	2000
5	Mago, Jenny Rose	1	.
6	Mago, Zaldy	2.5	1000
7	Magracia, Charlyn	1	600
8	Magramo, Merly	1.5	10000
9	Mariño, Gerlie	1	.
10	Mayo, Consolacion	2	10000
11	Mayo, Emperador	1	5000
12	Mayo, Jacinto Jr,	4	8000
13	Mazo, Melanie	7	2000
14	Mindo, Ulyses	0.5	500
15	Montesa, Marlon	1	1500
16	Mortel, Sharon	0.5	1100
17	Palabiano, Josie	1	500
18	Prado, Josephine	2	500
19	Roldan, Lenita	2	100
20	Roldan, Virgilio	2.5	.
21	Torres, Ederlyn	1	90
22	Torres, Leonilda	1	500
23	Villacruses, Pepita	4	600
	PROCESSORS		

	VILLAGE	FARM AREA	PRODUCTION VOLUME (brooms)
1	Burguete, Yolly		20000
2	Dela Vega, Merlinda		.
3	Miralles, Merlyn		10000
4	Race, Nilo		770
5	Veral, Benito		12000
DOÑA JUANA			
	FARMERS		
1	Bugarin, Remedios	0.5	.
2	Cordero, Melogine	0.25	.
3	dela Cruz, Leonila	2	.
4	Fadriquelan, Mariress	0.3	.
5	Gajolin, Lucila	1	600
6	Gallos, Adoracion	5	120
7	Gallos, Arceli	2	.
8	Gallos, Emilia	2	.
9	Gallos, Monico	1	.
10	Gallos, Natividad	0.5	.
11	Galos, Salvacion	0.1	.
12	Garachico, Enrigueta	0.5	.
13	Gatchalian, Maricel	1	.
14	Gonzales, Judito	3	.
15	Guarte,Delfin Jr.	0.5	.
16	Guro, Angeles	1	.
17	Guro, Arnel	2	.
18	Guro, Rodolfo/Elizabeth	5	400
19	Mabunga, Oscar	.	.
20	Mallen, Reynaldo	0.25	.
21	Malupa,Cesar	0.25	.
22	Manalon, Larry	0.8	.
23	Manasan, Mary Grace	1	.
24	Mangao, Rosafe	0.5	.
25	Mangat, Rolando	0	.
26	Mangera, Sandy	1	.
27	Manguera, isidro Jr.	0.5	.
28	Mansalay, Simplicio	2	.
29	Manzano, Penina	2	.
30	Manzano, Rudy	5	.
31	Manzano,Rogelio Jr.	0.25	.
32	Masungca,Ranilo	0.5	.
33	Mayo, Domingo	0.25	.

	VILLAGE	FARM AREA	PRODUCTION VOLUME (brooms)
34	Mayo, Ernesto	0.25	.
35	Mayo, Michael	0.25	.
36	Mayor, Clarito	0.5	.
37	Mayor, Rodolfu	1	.
38	Miralles, Cristy	0.5	.
39	Miralles, Janero	1	.
40	Miralles, silverio	0.5	.
41	Miran, Ronaldo	1	.
42	Montesa, Alice	5	.
43	Montesa, Alipio	2	.
44	Montesa, Alma	0.5	.
45	Montesa, Jennifer	0.5	.
46	Moral, Germelyn	1	.
47	Pupa, Enesita	2	.
48	Pupa, Helen	0.5	.
49	Pupa, Sonia	0.5	.
50	Recto, Marciano	0.5	.
51	Recto, Meldo	1	.
52	Recto, Ronnie	1	400
53	Rito, Charles	2	.
54	Rito, Jerry	0.5	.
55	Roldan, Crispin Jr.	1	.
56	Roldan, Elena	2.5	.
57	Salivio, Isaac	1	.
58	Sarabia, Danilo	0.5	.
59	Sarabia, Pablito	1	.
60	Tome, Casimira	1.5	.

	FARMERS & PROCESSORS		
1	Advincula, Ranlef	1.5	1000
2	Baldera,Sarah Jane	2	5000
3	Candor, Edmundo	1	1500
4	Clasyete, Renante	8	3500
5	Dela Cruz, Bernabe	2	1500
6	dela Cruz, Imelda	1.5	100
7	dela Cruz, Norma	7	1980
8	dela Cruz, Romeo	7	.
9	Fontelo, Elmer	1	300
10	Gallos, Loyd	4	150
11	Gallos, Mercedes	1	500
12	Gallos, Ramon Jr.	1	500

	VILLAGE	FARM AREA	PRODUCTION VOLUME (brooms)
13	Gallos, Teresita	0.5	400
14	Gallos, Wengie	1	500
15	Garachico, Nelmar	1	5000
16	Garachico, Nelson	0.25	300
17	Guro, Domingo	1	500
18	Guro, Florentino Jr.	2	1000
19	Ingreso,Marivic	1	200
20	Lapuz, Roger	3	.
21	Mabunga, Amada	1.2	5000
22	Mabunga, Arnel	3	.
23	Mabunga, Bruno Junior	0.5	500
24	Mabunga, Edelbeto	0.33	500
25	Mabunga, Eduardo	1	3000
26	Mabunga, Edwardo	2	.
27	Mabunga, Milagrosa	0.5	500
28	Mabunga, Romeo	0.5	3000
29	Mabunga, Rosemarie	0.5	1000
30	Mabunga, Rowena	3	60
31	Mabunga,Rolando	0.5	500
32	Magracia, Celso	0.5	4000
33	Magracia, Lucio	0.01	500
34	Magracia, Margarito	3	500
35	Magraia, Perlito	1	.
36	Manalon, Domingo Jr.	0.25	700
37	Manalon, Jolito*	1	100
38	Manalon, Josefina	0.5	3000
39	Manalon, Lina	1	100
40	Manalon, Pepito	1.5	.
41	Manalon, Renato	2	.
42	Manasan, Edwin	0.5	3000
43	Manasan, Rosa	1	500
44	Manato, Clarita	1	300
45	Manga, Castor	0.3	5000
46	Manga, Celedonia	1	1500
47	Manga, Frederick*	1.5	500
48	Mangao, Dreg	0.5	.
49	Mangao, Elena	1	1000
50	Mangao, Jury	4	1000
51	Mangao, Ruben	5	900
52	Mangera, Aida	1	10000
53	Marzonia, Vicente/ Nonita	2	.

	VILLAGE	FARM AREA	PRODUCTION VOLUME (brooms)
54	Masungca, Dolores	1.5	500
55	Masungca, Domingo	3	300
56	Masungca, Julie	2	500
57	Mayo, Francisco	6	12000
58	Mayo, Paulinia	5	.
59	Mayor, Teresita	1	400
60	Menes, Hesus	3	.
61	Menes, Rosemarie	3	1000
62	Mesana, Felonila	1	.
63	Mijares, Asuncion	1.5	3250
64	Montesa, Abad	0.5	500
65	Montesa, Donalyn	2	.
66	Montesa, Eleutirio	1	700
67	Montesa, Joemy/ Joy	3	100
68	Montesa, Juliana	1	800
69	Montesa, Jupiter	1	200
70	Montesa, Nilo	1	500
71	Montesa, Ronel	1	700
72	Montojo, Amadora	0.5	45
73	Mortel, Jaime/Roquata	2	600
74	Mortel, Janet	1.5	500
75	Mutya,Victoria	0	100
76	Pupa, Noe	0.5	5000
77	Recto, Elena	0.5	800
78	Recto, Rodel	0.5	1000
79	Reto, Efren	1.5	500
80	Rito, Samuel	1	500
81	Roldan, Joscuro	1	1000
82	Salivio, Erica	1	1000
83	Salivio, Gloria	1	300
84	Sarabia, Curren	0.5	1000
85	Sarabia, Rosana	0.03	500
86	Saravia, Dominador	1.2	75000
	PROCESSORS		
1	Bernardes, Concepcion		.
2	Manga, Angeles		80
3	Mangaya,Helen		1000
4	Molina, Tednard*		.
5	Montesa, Tessie		5000
6	Salivio Amparo*		1000

	VILLAGE	FARM AREA	PRODUCTION VOLUME (brooms)
VICTORIA			
	FARMERS		
1	Buhala, Alejandro	0.5	.
2	Gan, Vicente	2	.
3	Guntan, Edilberto	0.5	.
4	Guntan, Lioncia	1.5	.
5	Manucay, Ernina	0.5	.
	FARMERS AND PROCESSORS		
1	Guntan, Jenesis	1	5000
2	Guntan, Loida A.	1.5	500
MARI-SUR			
	FARMERS		
1	Antonio, Diony	1	.
2	Antonio, Iluminada	0.5	.
3	Antonio, Joseph Mariano	1	.
4	Antonio, Jovencio	0.5	.
5	Bautista, Noe	1	.
6	Dela Cruz, Artemio	2	.
7	Dela Cruz, Julian	0.75	.
8	Dela Cruz, Meryl Amancio	0.25	.
9	Dela Cruz, Normita	0.5	.
10	Espinida, Elvie	2	.
11	Fesarillo, Noviva	0.25	.
12	Gabayno, Mabini	0.5	.
13	Galindez, Ronnie	2	.
14	Gaspar, Erlinda	0.5	.
15	Gaspar, Gemma	0.25	.
16	Gaspar, Palermo	0.75	.
17	Ignacio, Annie	0.25	.
18	Magracia, Elsie	0.5	.
19	Mallorca, Arlen	2	.
20	Maneje, Rosita	0.25	.
21	Mortel, Catalina	3	.
22	Mortel, Jay-r	0.5	.
23	Mortel, Marcelo	1	.
24	Ravalo, Rebecca Galindez	1	.
25	Rioja, Christy	2	.

	VILLAGE	FARM AREA	PRODUCTION VOLUME (brooms)
	FARMERS AND PROCESSORS		
1	Claud, Annietta	2	500
2	Galicia, Daisy	1	200
3	Martinecio, Liberato, Jr.	0.5	100
4	Mortel, Samuel	2.5	300
PAGSANGAHAN			
	FARMERS		
1	Antaran, Berlito	0.5	.
2	Espelimbergo Jun	0.25	.
3	Fajilagot, Nelson	0.5	.
4	Fajilagot, Pablito	1	.
5	Falame, Antonio	2	.
6	Falsado, Edilito	0.5	.
7	Fernando, Dolphy	5.5	.
8	Fezarit, Pabes	1	.
9	Madeja, Luz	0.5	.
10	Madia, Joseph	0.5	.
11	Magro, Estrella	1	.
12	Magro, Mamelito	2	.
13	Magro, Nopreno	0.5	.
14	Mansalay, Catalina	0.5	.
15	Mariano, Ariel	3	.
16	Mariano, Arnel	0.5	.
17	Millares, Fred	1	.
18	Morada, Amado	2	.
19	Morada, Avelino	0.5	.
20	Morada, Ildefonso	1	.
21	Morada, Rocketson	1	.
22	Mortera Anecito	1.5	.
23	Mortera, Samparado	0.5	.
24	Moscoso ,Eric	1	.
25	Moscoso, Florencio	1	.
26	Moscoso, Miguelito	0.5	.
27	Rogon, Fernando Sr.	1	.
28	Romero, Patrocenio	3	.
29	Romero, Vicente	0.5	.
	FARMER AND PROCESSOR		
1	Madia, Vicente	1	1500

Appendix B

List of Buyers of Products

BUYERS OF PRODUCTS FROM BINONGAAN

1 Burguete, Yolly
2 Dela Cruz, Ailyn
3 Maduro, Maricar
4 Magramo, Merly
5 Manasan, Gingging
6 Mindo, Elizabeth
7 Mortel, Eddie
8 Mortel, Lelek
9 Salivio, Mayen
10 Saravia, Sabeth

BUYERS OF PRODUCTS FROM DOÑA JUANA

1 **Manasan, Toto***
2 Bautista, Marissa
3 Fajutag, Bingbing
4 Magay, Roging
5 **Manasan, Ruel***
6 Mangao, Bing
7 Mangaya, Ermund
8 Mangaya, Rogelio
9 Mindo, Elizabeth
10 Molina, Manet
11 Montesa, Elisa
12 Moreno, Inday
13 Sentino, Elisa

BUYERS OF PRODUCTS FROM MARI-SUR

1 Claud, Russel
2 Mortel, Eddie

*Big-time purchasers

ABOUT THE AUTHOR

EDDIE G. FETALVERO is an Associate Professor III of Biological Sciences at the College of Education, Romblon State University. He is currently working on his dissertation for a PhD degree in Education major in Biology Education at the University of the Philippines Open University. He began studying tiger grasses in a Village called Marigondon Norte in San Andres, Romblon in 2009 through a funding grant from the Department of Trade and Industry. In 2011, he expanded his research in the nearby seven adjoining villages with funding support from the Department of Science and Technology. Presently, he has just completed a University-funded research project which explores the ecological roles of tiger grass in sustaining the CALSANAG Watershed Forest Reserve.

www.ingramcontent.com/pod-product-compliance
Lightning Source LLC
Chambersburg PA
CBHW081842170526
45167CB00007B/2879

* 9 7 8 1 5 0 2 9 8 3 8 3 1 *